BBC

DOCTOR WHO

THE SECRET LIVES OF
MONSTERS

BBC

DOCTOR WHO

THE SECRET LIVES OF
MONSTERS

JUSTIN RICHARDS

Published in 2014 by:
Harper Design
An Imprint of HarperCollinsPublishers
195 Broadway
New York, NY 10007
Tel: (212) 207-7000
Fax: (212) 207-7654
harperdesign@harpercollins.com
www.harpercollins.com

Distributed throughout the world by
HarperCollinsPublishers
195 Broadway
New York, NY 10007

ISBN 978-0-06-234886-9

Library of Congress Control Number: 2014943185

Editorial director: Albert DePetrillo
Editorial manager: Lizzy Gaisford
Series consultant: Justin Richards
Design: Amazing 15
Original illustrations: Peter McKinstry
Production: Phil Spencer

Printed and bound in China by C&C Offset Printing Co., Ltd

1 2 3 4 5 6 7 8 9 10

ACKNOWLEDGEMENTS

With thanks to everyone involved in the making of Doctor Wh
in the past, the present and the future. And also, especially, to

Tom Spilsbury and Peter Ware of Doctor Who Magazine
Scott Handcock
Peter McKinstry
Neill Gorton and Kate Walshe of Millennium FX
Edward Russell
Steve Tribe
Mike Tucker and everyone at The Model Unit
Peter Tyler

BBC Books would like to thank the following for providing
photographs and for permission to reproduce copyright materia
While every effort has been made to trace and acknowledge al
copyright holders, we should like to apologise should there have
been any errors or omissions.

All images © BBC, except the following which are used with
kind permission: **Millennium FX** - pages 36 (top right), 38, 39
(middle left, top right), 120 (top), 130, 131 (top, bottom left)
162, 164, 165, 224 (top), 226, 231 (top, bottom right),
267, 268 (left), 272, 284, 285, 286, 287; **David Richardson**
- page 74; **Mike Tucker** - pages 66 (bottom), 67, 76 (top), 78
(bottom right), 232, 233; **Peter McKinstry** pages 2-3, 28-9,
87 (top), 102 (top right), 103, 121, 134, 154 (top left), 171
185, 204 (top), 220-1, 240-1, 262-3, 180, 281; all design
furniture © Shutterstock.

CONTENTS

WHEN THE DALEKS invaded Earth in 2009, it became suddenly and frighteningly clear that we are not alone in the universe. Of course, even before that we had the Battle of Canary Wharf, the 'Tenth Planet' incident, and the Mars Probe fiasco...

Throughout history, there have been rumours, legends, myths and stories of creatures visiting Earth from beyond the stars, from other worlds. Many of those stories are just that – fiction and speculation. But some of them are based on more than a grain of truth.

In this book we bring together the evidence that has emerged over the past few years and throw light into the darkest corners of the universe. We draw on information from the secretive UNIT organisation, and the even more shadowy Torchwood Institute. Many of the classified documents we refer to are only available now thanks to the untiring efforts of investigative journalists like James Stevens whose pioneering work has helped us to understand UNIT. Then there are the 'whistleblowers' – many of them anonymous – like Benjamin Nevis who have leaked some of the more comprehensive Torchwood reports.

What we learn from these various sources has helped us to build up a picture of our planet as a potential target for alien incursion and invasion. It has begun to reveal just how much the authorities have kept back – not just about past alien activity but also their knowledge of future events, the sources for which we cannot even begin to guess at.

But now at last the truth can be told, or at least as much of it as we can glean from the available evidence. The monsters are coming out of the shadows.

In this book you will find discursive articles about alien – and sometimes not so alien – monsters. You will see some of the original source materials and evidence reproduced here often for the first time.

The Cybermen

THE ARRIVAL OF A new planet in our solar system... Although memories seem to have faded and the authorities and scientific bodies now deny all knowledge of the event, in 1986 it was major news. There was speculation that the planet was a 'lost twin' of Earth, with the similarity between the land masses on both planets noted by astronomers. But then, the new planet exploded in a dazzling display of light and colour. Most people on our own planet witnessed the explosion or its after-effects.

What most people did not know was that the planet – named Mondas – was inhabited. The people of Mondas, anticipating the fate of their own world, tried to destroy ours so that Mondas would survive. They sent raiding parties to key military and political centres to take control, and almost succeeded in using a doomsday weapon – the Z-Bomb – to destroy the Earth. Despite the fact their own planet resembled ours, the people who came here from it were very different. They had been exactly like us once, but now they were Cybermen.

Discovering that their lifespans were contracting and their bodies becoming weaker with every generation, the people of Mondas had come up with an extreme solution. As on Earth, spare part and transplant surgery was commonplace. But they followed it through to its logical conclusion. With the human body entirely replaced by plastic and metal, with brains enhanced by digital technology and all irrationality and emotion removed, logic was all they had left. Logic, and the impulse to survive no matter what.

The creatures who came to destroy the Earth were more machine than human. Each had the strength of ten men, and the ability to survive in the airless vacuum of space. But they had no feelings, no pity, no emotion. No humanity.

Even those few people who did witness the attack of the Cybermen were unaware that this was not the first time the creatures had attempted to conquer our planet. Or that they would be back...

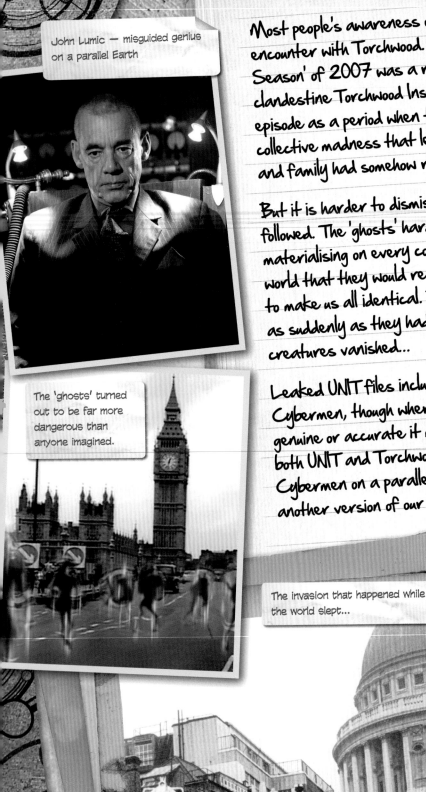

John Lumic — misguided genius on a parallel Earth

The 'ghosts' turned out to be far more dangerous than anyone imagined.

The invasion that happened while the world slept...

Most people's awareness of the Cybermen comes from their encounter with Torchwood. It is now well documented that the 'Ghost Season' of 2007 was a result of experiments carried out by the clandestine Torchwood Institute. Many now write off that whole episode as a period when the world was possessed by a kind of collective madness that led to people believing their departed friends and family had somehow returned as ethereal, insubstantial 'ghosts'.

But it is harder to dismiss the invasion of the Cybermen that followed. The 'ghosts' hardened into men of steel and plastic, materialising on every continent – and even broadcasting to the world that they would remove our fear, sex, class, colour and creed to make us all identical. To make us Cybermen. And then, just as suddenly as they had appeared, and just as mysteriously, the creatures vanished...

Leaked UNIT files include hints and details of future battles with Cybermen, though where this information derives from and how genuine or accurate it might be is not recorded. Of particular note, in both UNIT and Torchwood files, are references to the development of Cybermen on a parallel Earth – not a 'twin' planet like Mondas, but another version of our own world.

Here the Cybermen were developed less out of necessity and more to exploit available technology. That said, the man behind the creation of the Cybermen, John Lumic, had a very personal stake in their development as his own body was wasting away and cybernetic conversion must have represented his only real hope of survival. The fate of the people of Mondas was played out on a very personal level in Lumic's own plight. Using the resources and technology of his multinational corporation Cybus Industries, Lumic developed the Cybermen, and unleashed them on his own world.

From there, these Cybermen somehow found a way to escape through a 'void' to our own universe. It is Cybermen from Lumic's world that appeared as an army of ghosts, and which seem to have arrived in London in 1851. The origin of the Cybermen UNIT faced in the 1970s is less clear. Again, they were helped by a multinational corporation, in this case International Electromatics, the company behind the micro-monolithic circuit that was almost pervasive in the technology of the time. But the company's Chief Executive, the charismatic and brilliant Tobias Vaughn, was in league with the Cybermen.

According to UNIT, the Cybermen did actually invade and take control of large parts of London before they were destroyed. The reason no one was aware of this is because all IE technology was sabotaged and emitted a hypnotic wave that put the world's population to sleep. When the invasion had been averted, the wave stopped and the world awoke – having lost several whole days without anyone noticing...

The Cybermen developed by Cybus Industries (note the 'Cybus' logo embossed on the chest)

The true identity of the 'ghosts'

a sword from a display on the Reverend's wall. So armed, the redoubtable Mr Smith turned at once to do battle with the metal demon. But even as he struck out at the creature, a second, identical to the first, appeared. Smith was driven back, his sword-blows as nothing to these demonic beasts of metal.

The Doctor himself, meanwhile, remained in a fugue. Smith's words and the memories stirred from finding the strange apparatus had him all befuddled. Again, it was as if he was two people within a single body. Two minds vying for control. And yet, there was a memory that stirred which might help.

As the air resounded to the noise of sword on steel, blade on metal man, the Doctor hunted desperately for the infostamp which Smith had been forced to discard when the Cybermen attacked. At last, he spied the device, and gathered it up with an urgency born of desperation. If he could but recall how he had tampered with the device before, then perhaps all was not yet lost.

But the Cybermen were driving Smith back up the stairs, despite his bravery and skilled swordsmanship. As they reached the upper floor, it seemed that all was lost. The two metal men closed in, as they thought, for the kill. They were to be disappointed.

For at that moment, the Doctor finally managed to open the capped end of the infostamp device. A lightning fire was emitted, blazing across the room to strike the Cybermen. With a noise that cracked the ears and echoed round the building, their metal heads exploded into fragments. Smith stepped back, amazed. The Doctor calmly returned the cap to the device's terminus. The two figures, devoid of heads, crumpled lifeless to the floor.

For several moments, all was silence. Then the urgency of their situation bore down upon the two comrades in arms, and

There rose up apparently from beneath the river itself a hideous great figure, all made, it seemed of metal. Many in the crowds that soon gathered to gain a view of this spectacle believed it to be some festive celebration. Some even claimed to have seen the mighty creature displayed at the Great Exhibition earlier in the year. But then the mechanical monstrosity did shoot fire from its arms, to the cause of great destruction on the ground below.

Imagine our surprise and fear when another extraordinary sight appeared in the night sky. I have heard tell of these hot-air balloons, but this was the first I had witnessed in person. Dwarfed by the size of the enormous metal figure, the balloon with passenger in a basket slung below seemed at first to seek

parlay with the mechanical beast. When this, as it seemed, failed, then battle ensued. To the amazement, not to say delight, of those of us assembled, the balloon spat fire at its opponent, which did stumble and then vanish in a burst of light and fire that was quite spectacular to behold and which drew cheers from us all.

Were it not for the damage to property and the loss of life, which was not inconsiderable, I would say this was the most amazing Christmas spectacle London has beheld.

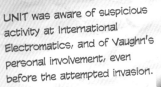

It is now well over a year since these objects were first sighted. While fighter planes have been sent to intercept, the unidentified craft have continued to elude them. Contact is lost, in most cases, over south east England, in an area close to the International Electromatics research facility.

A UNIT operative sent to investigate failed to return. Surveillance on the IE headquarters building in London has been stepped up. As a result, we can confirm that several people who entered the building have never emerged, or at least not by any conventional exit. The rail link between IE facilities remains a possible egress.

While this evidence remains circumstantial, information from other sources suggests that IE is indeed implicated. The disappearance of Professor Watkins has been linked directly to Tobias Vaughn himself, and to his Head of Security, Packer.

UNIT operations are ongoing, and further reports will be forthcoming as and when more information becomes available.

A photograph developed from the hidden camera carried by undercover police officer Vincent Russell, who disappeared in 1985. His camera was found intact in a London sewer.

GENERAL CUTLER: The expected attack, sir. They've been sighted in force.

SECRETARY WIGNER: Yes, I know. We've just got reports. They are coming in from all over the world. And to make matters worse, the energy drain is increasing rapidly. Cutler, you must hold on as best you can.

>> BREAKUP – UNINTELLIGIBLE <<

GENERAL CUTLER: The Z-bomb, sir. Mounted in the warhead of a Demeter rocket and fired at Mondas, it could destroy it.

SECRETARY WIGNER: We can't take the risk.

>> BREAKUP – UNINTELLIGIBLE <<

GENERAL CUTLER: But there isn't time for consultation. This is an emergency!

A fragment of an intercepted communication between General Cutler, officer commanding Snowcap Base in Antarctica, and International Space Control headquarters in Geneva.

These photos from 1988 appear in the UNIT files without explanation. The man has been identified as Hans De Flores, a Nazi SS officer who fled to South America in 1945. This is the last confirmed sighting of the war criminal, who was convicted at Nuremburg in absentia.

And there was at that time, living in Windsor, a witch. Though but she went by the name of Peinforte, she was known to be possessed of devils and did make spells and commit foul murther by means thereof.

She did forge a statue of living metal which did imitate her aspect and was much admired. This statue, some say, was so close to life that it possessed the power of speech and did utter strange and powerful words. It came to pass at last that a mighty wizard or warlock did cast the statue out of this world and into the stars, there to wander for all eternity or till it should again return to its earthly home and wreak great damage upon humanity.

In the Year of our Lord 1638 was Lady Peinforte also taken as if by magick along with a ruffian klept Richard though this ruffian did later return and was agreed by all to be of character much reformed.

Photographed in Windsor in 1988, this woman bears an uncanny resemblance to the seventeenth-century Lady Peinforte. The man has not been identified.

Lemon VITEX LITE

PETE TYLER says...

'Trust Me on Th[is]'

Pete Tyler

Cybus diversified into many industries — including food and drink.

Pete Tyler says...

'Trust Me on This'

NEW CHERRY VITEX LITE

Pete Tyler

An invitation to Jacqueline Tyler's 39th (some reports say 40th) birthday party. Among the guests was the President of Great Britain. He was killed when the first Cybermen attacked.

Pete Tyler cordially invites you to celebrate the birthday of his wife

Jacqueline Tyler

On 1st February at the Tyler Residence, from 8pm

Black Tie   RSVP

Pete says: "It'll Be A Party You Will Never Forget – Trust Me On This!"

READY TO UPGRADE?

Then step up to the latest release from the Cybus Network*

NEWS • SPORTS • WEATHER • TV GUIDE • LOTTERY NUMBERS

Whatever you're interested in – Cybus Network Can provide It

Sign up for the Daily Download
Direct to your Cybus SmartPhone
Or Direct to YOU

As with all Cybus Industries products, complete compatibility is guaranteed

* A Cybus Industries company

Cybus Industries controlled almost all network access and communication on the parallel Earth, owning many subsidiary companies including International Electromatics, Vitex, IE24, Cybus Finance, Cybus Properties, and Cybus Network.

© COPYRIGHT CYBUS INDUSTRIES
ALL RIGHTS RESERVED CODE : PJM-20073

Photographs taken by guests at Jackie Tyler's party.

In memory of the sacrifice made by Sir Robert MacLeish, and in memory of our ever-loving husband and consort Albert we, Victoria Regina, in the Year of Our Lord 1879, make the following decree.

That an Institute shall be established and named after Torchwood House, ancestral home of the MacLeish family. The purpose of this Torchwood Institute shall be to defend Great Britain and her Empire from dangers beyond imagination. It shall investigate all strange happenings and oppose such as are deemed threats to our borders on all sides.

And in addition thereto the Torchwood Institute shall keep vigil for the return of The Doctor, dubbed by us Sir Doctor of Tardis as he had done great service before his banishment, exiled from our Empire never to return. He remains and ever shall be an enemy of the Crown.

This order and decree herein made in perpetuity by

Victoria R.

Victoria, by the Grace of God, of the United Kingdom of Great Britain and Ireland and of Her other Realms and Territories Queen, Defender of the Faith.

Images retrieved from the high-resolution security surveillance systems at Torchwood Tower following the Battle of Canary Wharf in 2007

The short but chilling message broadcast by the Cybermen

"CYBERMEN NOW OCCUPY EVERY LAND MASS ON THIS PLANET, BUT YOU NEED NOT FEAR. CYBERMEN WILL REMOVE FEAR. CYBERMEN WILL REMOVE SEX AND CLASS AND COLOUR AND CREED. YOU WILL BECOME IDENTICAL. YOU WILL BECOME LIKE US."

Extract from a 2007 report by Yvonne Hartman, Chief executive Officer of the Torchwood Institute. She was reported lost, believed killed, in the Battle of Canary Wharf.

CONFIDENTIAL

TORCHWOOD
INSTITUTE

Torchwood's Research and Development Department continues to suggest ways of analysing the 'Sphere' but so far it has resisted all of them. The new spectrometer failed to register any activity or heat from the object, which remains inert.

Our chief analyst, Rajesh Singh, maintains that he has a 'feeling' the Sphere is watching him. But of course this cannot be quantified, any more than the feeling it gives us all that we just want to run away and hide from it.

All we can be certain of, despite the evidence of our own eyes, is that the Sphere does not exist. It weighs nothing, emits no heat or radiation, does not age, and has no atomic mass.

We do however remain convinced that the Sphere is somehow linked to the Ghost Shift. We have continued to monitor drops in power levels during the most recent Ghost Shifts, but have been unable to correlate these to any Sphere activity. Not least because there is no Sphere activity. Whatever it is, or whatever it contains, remains a mystery that will perhaps never be solved.

CONFIDENTIAL

Fragment of a radio conversation between Hobson, commanding Moonbase, and ISC HQ. The virus was later identified as Neurotrope-X, developed by the Cybermen.

Photographs beamed back to Earth from the Lunar Relief Shuttle before it lost control and plunged into the sun.

ISC: International Space Control Headquarters Earth calling Weather Control Moon. Come in please.

HOBSON: Hobson here.

ISC: What's happening up there? The hurricane you were guiding is now 45 degrees off course. It's threatening Hawaii.

HOBSON: One of my men was taken ill. We're operational again now.

ISC: The controller would like to know the cause of the illness.

HOBSON: Yes, so would we. We've now got three men down with this mystery virus in the past few hours, including the doctor.

ISC: One moment please. Stand by for further instructions.

... BREAK ...

ISC: Your instructions are to send blood samples back to Earth for investigation.

HOBSON: Well, when can we do that? The next shuttle rocket's not for another month.

ISC: Then they must be put on that rocket. In the meantime the Moonbase is to be put in quarantine.

THE CYBERMEN

THE MYSTERY OF THE *SILVER CARRIER* IS PERHAPS EVEN MORE PERPLEXING THAT THAT OF THE MORDEE EXPEDITION AND OTHER LOST SHIPS. WHAT IS SO BEWILDERING IN THIS CASE, IS THAT THE *SILVER CARRIER* TURNED UP AGAIN.

A PHOENIX MARK IV SERVICE AND SUPPLY SHIP, THE *SILVER CARRIER* WAS EN ROUTE FOR STATION FIVE WHEN SHE WAS REPORTED LOST, JUST 7 MILLION MILES OUT. A THOROUGH SEARCH OF THE AREA FAILED TO LOCATE THE VESSEL, AND ALSO FOUND NO WRECKAGE OR DEBRIS. INITIALLY, HER DISAPPEARANCE WAS PUT DOWN TO A METEOR STRIKE, AND SHE WAS DECLARED LOST, TOGETHER WITH HER CREW OF TWO.

BUT THEN, NINE WEEKS LATER, THE *SILVER CARRIER* WAS FOUND. DESPITE HAVING FUEL FOR APPROXIMATELY 20 MILLION MILES WHEN SHE WAS LOST, SHE REAPPEARED 87 MILLION MILES AWAY — IN THE VICINITY OF STATION THREE. NUMEROUS THEORIES HAVE BEEN PUT FORWARD AS TO HOW THIS COULD HAVE HAPPENED. MOST PLAUSIBLE IS THAT THE SHIP DRIFTED, BUT AN UNPOWERED FLIGHT OF SUCH DISTANCE AND SPEED HAS NEVER BEFORE OR SINCE BEEN RECORDED.

LITTLE WAS MADE OF THE *SILVER CARRIER* MYSTERY AT THE TIME. THE EVENT WAS LARGELY OVERSHADOWED BY THE TRAGIC LOSS OF LIFE ON STATION THREE ITSELF, SUFFERED DURING A FREAK METEORITE SHOWER.

The description of the Silver Carrier mystery from Alnecht Kohl's *Great Mysteries of the Universe*. The pictures show the interior of the *Silver Carrier*, taken soon after her construction.

Notes taken by Bill Duggan, one of the crew of Station Three killed in the alleged meteorite strike. Is he describing a Cybermat? Were the Cybermen to blame for events on Station Three?

A standard Servo Robot of the type used on board the Silver Carrier

Billy Bug — found close to X-Ray laser equipment. Unlike any life form I've seen before. Made of metal!!

About 3°cms long

2 bulbous eyes and segmented tail.

Ridge of spines down back

Legs or feelers (?). No sign of a mouth.

Has somehow managed to corrode the Bernalium stock!!

Dig Deep for the Truth

Professor Parry briefly describes his expedition to Telos in his memoirs *Dig Deep for the Truth*. But much about the expedition remains unexplained.

And so, with the hatch finally open, we descended to the lower levels. The technology here was decidedly more sophisticated, far in advance of the controls in the entrance hall some of which dated back to the Cybermen's earliest dynasties.

The air became predictably but decidedly cooler as we found ourselves in a vast cavern beneath the city. And here we found them — the tombs of the Cybermen. The creatures themselves were perfectly preserved within a vast honeycomb of frozen cubicles. A moment unique in archaeology.

The Doctor spoke of their great evil being locked away for all eternity. Klieg compared them to bees awaiting the signal to arise from their winter sleep. Up to that moment, I had never wondered why Klieg and Kaftan should be so interested in financing my expedition. I assumed they shared my own enthusiasm for unearthing the mysteries of the past, my passion for discovery. I was soon to discover how naïve that assumption was.

But for the moment, I stared in awe at the frozen tombs, proud to be the leader of an expedition that had made such a discovery — that had at last unearthed the secrets of the Cybermen. Just minutes later I was wishing I had never come to Telos…

While regrettable, the loss of the freighter was unavoidable. Although, we can take some comfort from the fact that the cargo was not as stated on the manifest, but actually an army of inert Cybermen, the loss of human life is lamentable. Captain Briggs and her second in command Berger managed to evacuate the Freighter, together with Lieutenant Scott and several of his troopers. But most of the crew - including First Officer Ringway (see above) - was lost.

It is worth stressing to Galactic Insurance that had the freighter not been diverted from its collision course with Earth, disappearing from all tracking systems (presumably vaporised), the consequences would have been catastrophic and the company's liabilities far in excess of the relatively modest sum we are now claiming in recompense.

We attach statements from the Earth Security Forces Directorate, verifying the information we have already provided, and hope that in lieu of wreckage and debris this will suffice to prove to Galactic that the freighter has indeed been destroyed and that our claim is valid.

Extract from an insurance claim for a space freighter apparently destroyed by the Cybermen. The picture of the freighter's bridge shows Captain Briggs (on the right), Berger (left) and an unidentified crew member (centre).

Pictures extracted from the freighter's security systems, which were backed up to the nearest comms satellite before the freighter vanished.

03. CYBER-WARS

Targus Migarian's *A History of the Cyber Wars* is the definitive historical account of humanity's battle against the Cybermen.

EVEN BETWEEN THE great Cyber Wars, there were sporadic and isolated instances of Cyber incursion. Two incidents in particular are worthy of note – the Nerva Beacon plague, and the destruction of Hedgewick's World.

In the late twenty-ninth century, a large asteroid was captured by Jupiter's gravity and went into orbit round the planet. Since it was not yet marked on all star charts, Nerva Beacon was placed nearby to warn ships away from Jupiter's newest moon. What was not apparent until later was that the asteroid – designated Neo Phobus – was in fact the largest surviving fragment of Voga, the fabled Planet of Gold.

Desperate to destroy the last remnants of Voga before launching a new offensive against humanity, a surviving group of Cybermen used Cybermats carrying a variant of the Neurotrope-X virus to infect the crew of Nerva. Taking control of the beacon they then planned to destroy Voga using Cyberbombs. But what they did not know was that their human ally on the beacon was in fact working for the Vogans and their Cybership was destroyed by a Vogan missile before their plan could be carried out.

Events on Hedgewick's World are less well documented. What is known is that Hedgewick bought the planet cheap after it was devastated in the Cyber Wars, and constructed a huge amusement park on the planet. But it seems that Cyber casualties from the Cyber Wars of the 250th Millennium were secretly hidden beneath the surface of the planet. The Cybermen took humans from the amusement park, using them as spare parts to rebuild and upgrade their growing army. This activity all but ceased when the amusement park was closed down and abandoned.

Following the drastic but effective destruction of the Tiberion Spiral Galaxy to halt the advancing Cyber Armies, it seemed that the creatures had finally been wiped out. But the army on Hedgewick's World was reactivated. Millions of upgraded Cybermen prepared for battle. Again drastic action was needed, and Hedgewick's World was utterly destroyed by the military unit based there. Rumours that Emperor Ludens Nimrod Kendrick called Longstaff, the forty-first, defender of humanity, and Imperator of Known Space was somehow involved in the incident have never been confirmed.

In its heyday, Hedgewick's World was said to be the biggest and best amusement park ever created.

HEDGEWICKS WORLD

BOATING LAKE

VOLCANO RIDE

HOTEL

SAND DUNES

FOLLY RIDE

PORT

NATTY LONGSHOE'S CASTLE

MUSHROOM WORLD

SERVICE AREA

HOTEL

FIREBALL

VIEWING TOWER

GARDEN

THE GIANT'S CAULDRON

EYEBALL TERROR

FOODHALL

MEDIA HUB

SHOP

VIEWING TOWER

TOPSY TURVY RIDE

MOTEL

MAIN ENTRANCE

HEDGEWICK

CORAL KINGDOM

SHOP

BAR

THE JOLLY BELLY

One of the attractions of Hedgewick's World, ironically, was this chess-playing replica Cyberman, part of Webley's World of Wonders.

...ION VIA THE BEACH
...ION VIA MEDIA HUB
...ION VIA THE JOLLY BEAN
...ION VIA EYEBALL TERROR
...ION VIA CORAL KINGDOM
...ION VIA VIEWING TOWER

A Cyber hibernation chamber of the sort believed to have been hidden beneath Hedgewick's World.

The prime motivation for any race is to survive – and the Cybermen are no exception. Everything they do is motivated by the desire to perpetuate their race. Obviously they do not reproduce biologically, so the survival instinct emerges as a need to 'upgrade' and convert other humans, turning them into Cybermen.

There are documented instances of partial conversion, the most obvious being Tobias Vaughn, whose body was apparently cybernetic, although he retained his own head and brain. Professor Watkins recalls in his diary an instance when Vaughn goaded him into shooting the man – the bullets having no effect, to Vaughn's evident amusement. It seems likely that several IE employees were also 'augmented' with cyber technology.

A more limited form of control can be exerted by means of headsets that emit a hypnotic wave, perhaps similar to the wave that disabled the world during the Cybermen's attempted invasion in the 1970s. This control system seems to have been used by the Cybermen who attacked the Moonbase, as well as to control Torchwood operatives at their Canary Wharf headquarters in 2007.

Both UNIT and Torchwood files contain unpleasant and graphic details of the cyber-conversion process as carried out partially on, for example, Toberman in the Cybermen's tombs on the planet Telos and on the mercenary Lytton and others. Luckily, there was not time for many people to be converted during the 2007 invasion.

A Cybershade of the type that terrorised Victorian London

The Cybernisation process

Cybermites — the ultimate development of the Cybermat?

But it is not just people that the Cybermen convert. Evidence from the 1851 incident suggests that 'Cybershades' were created out of animals in London. These feral creatures were a combination or organic and contemporary industrial materials, and designed to be less conspicuous in Victorian London than the Cybermen themselves. They were used primarily for reconnaissance, but also to instil fear into the local population.

Smaller and more insidious are the Cybermats. Small, metallic creatures not unlike rodents, the Cybermats can home in on human brainwaves to attack. On Nerva Beacon, the Cybermats disseminated a plague that killed most of the crew – possibly a variation of the Neurotrope-X with which the Cybermen attacked the Moonbase. Bizarrely, a creature resembling a Cybermat was even reported in a department store in Colchester.

Much smaller, tiny Cybermites are the smallest evolution of Cybermat technology and can upgrade any humans they come into contact with, converting them into Cybermen. At the opposite end of the scale is the CyberKing – a huge Cyber factory and Dreadnought-class warship capable of converting millions of humans into Cybermen.

Creatures based on technology, the Cybermen are constantly upgrading themselves – always for increased efficiency rather than aesthetics. They seem able to exploit digital systems in ways we have not yet even thought of. For example, if a Cyberleader is destroyed, its memories and experience are automatically downloaded into another Cyberman which takes over the role.

Cyber weaponry is also evolving at a pace. Different versions of Cybermen are able to transmit a lethal electric current through their hands, and some can emit a hypnotic wave. Some models have weaponry built in – part of the head, the chest unit, or emerging from the wrist – while others make use of more conventional guns.

For all their upgraded technology, the Cybermen do have weaknesses. They are susceptible to the extreme effects of gravity, as seen on the Moon. During the 1986 invasion from Mondas, Cybermen were destroyed by radiation, and were dependent on their home planet for power and even life. Solvents have been used to dissolve vital plastic components particularly on versions with exposed chest units (that seem to replace the heart and lungs).

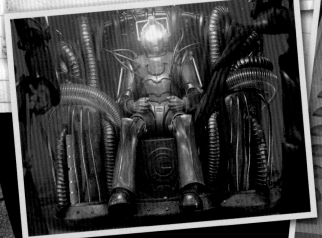

Inhabitants of Voga, the planet of gold

Gold can also be used against the Cybermen. Being non-corrodible, powdered or liquid gold plates the creature's breathing apparatus, in effect suffocating them. In the Cyber Wars of the future, the Glitter Gun will be a decisive weapon, all but wiping out the creatures using supplies of gold from the planet Voga where it is abundant.

Perhaps ironically, for a race that has removed all feelings from their brains, the Cybermen are also vulnerable to extreme emotions. The Cybermen created by John Lumic were defeated when their emotional inhibitors were deactivated – exposing the creatures to sudden, violent emotions that destroyed them. Similarly, Tobias Vaughn developed Professor Watkins' teaching machine, the Cerebratron Mentor, into a weapon. Injecting fear directly into a Cyberman's systems drove it mad, while more concentrated bursts of extreme emotion destroyed other Cybermen.

BEHIND THE SCENES

The Cybermen were the creation of Dr Kit Pedler, who became for a while unofficial scientific adviser to *Doctor Who*. Pedler was a medical researcher when he was recommended to script editor Gerry Davis as someone who might be useful. He and Davis worked closely together to develop ideas – including the Cybermen.

Pedler's great fear was 'dehumanising medicine.' He foresaw a time when spare part surgery had reached the stage where it was commonplace, possibly even cosmetic. There would come a point where it was impossible to tell how much of the original human being remained. Such creatures, he reasoned, would be motivated by pure logic coupled with the overriding desire to survive. They would sacrifice their entire bodies and their minds in the quest for immortality… From this fear, the Cybermen were born.

The job of designing the Cybermen from these descriptions fell to costume supervisor Sandra Reid. While the look of the Cybermen would change and evolve with almost every story in which they appeared, the initial design included the main elements that still make the Cybermen recognisable: the blank mask-like face, 'handles' connected to a light in the head, cables and rods to enhance the limbs, and the large chest unit.

The voices of these early Cybermen, however, were markedly different from their successors. In the script, the Cyberman voice was described as 'flat… hard in tone.' But voice artist Roy Skelton came up with a stilted, eerily mechanical-sounding

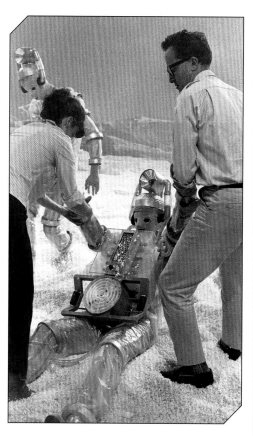

THE FIRST DOCTOR

THE TENTH PLANET
Written by Kit Pedler and Gerry Davis
Directed by Derek Martinus
First Broadcast: 8-29 October 1966

THE SECOND DOCTOR

THE MOONBASE
Written by Kit Pedler
Directed by Morris Barry
First Broadcast: 11 February-4 March 1967

THE TOMB OF THE CYBERMEN
Written by Kit Pedler and Gerry Davis
Directed by Morris Barry
First Broadcast: 2-23 September 1967

THE WHEEL IN SPACE
Written by David Whitaker (from a story by Kit Pedler)
Directed by Tristan de Vere Cole
First Broadcast: 27 April-1 June 1968

THE INVASION
Written by Derrick Sherwin (from a story by Kit Pedler)
Directed by Douglas Camfield
First Broadcast: 2 November-21 December 1968

THE FOURTH DOCTOR

REVENGE OF THE CYBERMEN
Written by Gerry Davis
Directed by Michael E. Briant
First Broadcast: 19 April-10 May 1975

THE TENTH PLANET 1
... the sleeve on the arm of one of them slips back. Instead of flesh there is a transparent "arm-shaped" forearm containing shining rods and lights. There is a normal hand at the end of it. A close-up of one of their heads reveals a metal plate running between centre hair line front and occiput...

voice with the pitch, intonation and inflection all different from normal human speech. While it was distinctive, however, this did not continue through to other stories and by the time they returned in *The Moonbase*, the Cybermen had become more electronic sounding.

Kit Pedler and Gerry Davis continued to write Cybermen scripts together – collaborating on *The Moonbase* and *The Tomb of the Cybermen*, and Pedler went on to contribute story ideas for *The Wheel in Space* and *The Invasion*. Later the two would create the landmark BBC ecological thriller series *Doomwatch*, also writing three related novels.

BEHIND THE SCENES

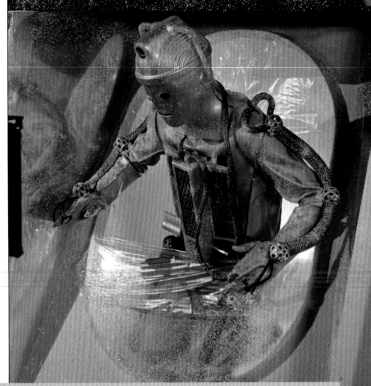

THE INVASION
The cocoon begins to pulsate and writhe, the cyberman inside beginning to come to life. The pulsating sound is now almost a shriek as the cocoon splits and the terrible silver giant cyberman appears.
We zoom in close on him as the brilliant white light illuminates his massive form.)

With more money and a new vinyl-backed material available for the Cybermen's next appearance, *The Moonbase* saw a complete redesign. The result retained the chest unit and handles, and was far closer to the Cybermen as we know them today.

The face, rather than being covered with cloth, was now a blank helmet not unlike a skull with its gaping mouth and sightless eyes. Vacuum cleaner tubes and plastic practice golf balls formed the hydraulic exoskeleton of the creatures.

The Cybermen were back again just six months later in *The Tomb of the Cybermen*. The costumes were almost unchanged, though extra pipes were added below the chest unit. This story also introduced the Cyber Controller – who wore the same basic suit, but devoid of chest unit. The Controller's head was less metallic, without handles and with an enlarged cranium.

The Cybermats were also introduced in *The*

Tomb of the Cybermen, and reappeared – in a slightly redesigned form for the next story, *The Wheel in Space*.

Only two Cybermen appeared for the bulk of *The Wheel in Space*. These were new costumes, designed by Martin Baugh and based on a wetsuit painted silver with more a streamlined exoskeleton of metal rods. The heads were similar to the previous version, adapted from existing helmets, but introduced the 'tear-drop' eyes. As well as bringing a hint of pathos to the design, the small holes added to the corners of the eyes (and in this case below the mouth) were to increase ventilation as the suits were extremely hot to wear. The chest units were also reused, but worn upside-down (except for one additional Cyberman in the final episode).

For *The Invasion*, the Cybermen were again redesigned, this time by Bobi Bartlett. A wetsuit once more formed the basic body, but the chest

units and hydraulics were more streamlined. The biggest change was the Cybermen's heads. These were made from lightweight fibreglass which meant they were easier for the actors to balance on their shoulders than previous, heavier versions. The lower part of the mask extended down the neck, and was covered by the top of the wetsuit, zipped up (sometimes rather obviously) to under the chin. The aperture at the top of the Cyberman's head was left empty, although the original intention had been to fit a working light into this socket.

The 'tear-drop' cut-outs in the corners of the eyes were kept from the previous design and the eye and mouth sockets were covered with thin gauze to hide the actor inside. The 'ear muffs' (which were then retained throughout the classic series Cyberman stories) were added for *The Invasion*. Partly these were to bulk out the head and give the impression of a powerful creature, and partly to allow more space inside the helmet both for the actor's head and for ventilation.

BEHIND THE SCENES

It was nearly seven years before the Cybermen returned properly to *Doctor Who*. For *Revenge of the Cybermen*, four costumes were created by external contractor Allister Bowtell. The main body was again a wetsuit, with new heads moulded from fibreglass and based very much on the design of *The Invasion*. The chest units were closer to those seen in *The Moonbase*, and detailed with parts from old television sets. Again, vacuum cleaner hoses were used for the hydraulic pipes.

For the first time, the Cyberleader was distinguished from other Cybermen by black detailing on his helmet. It was also the first time that the actors playing the Cybermen spoke their lines.

The Cybermat that appeared in this story was very different to previous versions and much larger.

There was another seven-year gap before the Cybermen returned in *Earthshock*. Again, the Cybermen were completely redesigned, this time by Dinah Collin. Built by the Imagineering company, the heads were similar to the previous design but with more detailing. The basic costume was built around a military flight suit, which already had tubes and pipes woven into it. The chest unit became more integrated, also forming a collar.

The mouthpiece was transparent, so that the actor's chin – painted silver and covered with thin plastic – could be seen inside giving a hint that something organic survives. But the transparent section tended to steam up. Also, the heat inside the costume meant that the battery pack for the actor's microphone, which was taped to the top of the helmet, slipped down into view.

The same Cyber costumes were reused, with mouthpieces painted silver, for *The Five Doctors* and again for *Attack of the Cybermen*. For this story, the Cybermen guarding their secret base in the London sewers were painted black as camouflage. The Cyber Controller reappeared, but as a standard costume with a simple, silver, enlarged cranium.

For *Silver Nemesis*, broadcast to coincide with *Doctor Who*'s 25th anniversary, the Cybermen

SIGNIFICANT SCREEN ENCOUNTERS

THE ELEVENTH DOCTOR

THE PANDORICA OPENS / THE BIG BANG
Written by Steven Moffat
Directed by Toby Haynes
First Broadcast: 19–26 June 2010

CLOSING TIME
Written by Gareth Roberts
Directed by Steve Hughes
First Broadcast: 24 September 2011

NIGHTMARE IN SILVER
Written by Neil Gaiman
Directed by Stephen Woolfenden
First Broadcast: 11 May 2013

```
                    ATTACK OF THE CYBERMEN
         SEVERAL CYBERMEN GO ABOUT THEIR ROUTINE DUTIES.
  ANOTHER CYBERMAN IS BEFORE A CONSOLE. HE TURNS AND ADDRESSES THE CYBER CONTROLLER.
  (Note: The controller differs from a regular cyberman in as much as that his head is
                        larger and domed shape.
  The controller is also taller and lacks much of the pipework that adorns the average
  cyberman. Otherwise he possesses the same empty, emotionless voice and manner of the
                              cybermen)
```

44.06.79C.28   CYBERMAN "OH DEAR"

45.06.79C.29   CYBERMAN FALLS HOLDING ONTO CUT HALF OF
ROPE LADDER.
POWERSTATION EXPLODING IN BACKROUND.

were again redesigned, by Richard Croft, though they stayed very close to the previous version. Perhaps the most noticeable difference was that the helmets and chest units were made to appear more highly polished using a new chroming process. An unfortunate side effect to this was that over time, the chrome oxidised, turning gold in colour.

BEHIND THE SCENES

The redesign of the Cybermen for their two-part reintroduction in 2006 – *Rise of the Cybermen* and *The Age of Steel* – was a month-long process. The *Doctor Who* Art Department produced a number of 'concept' paintings that showed a variety of different designs

The job of finalising the new design and creating the Cybermen themselves was given to the Millennium FX company, run by Neill Gorton, who had already produced some startling creature and prosthetic designs for the series, starting with the Autons in *Rose*.

For the new design, it was decided that the Cybermen should be able to turn their heads independently of the rest of the body, and also if possible tilt them up and down. The final head also included blue LED lighting which was illuminated when the Cyberman spoke – the voice being supplied by actor Nicholas Briggs, who also provided voices for the Daleks.

After creating various drawings and clay models of possible designs for the new Cyber head, a maquette – a small model – of the final design for

the Cyberman was made from clay. With the full design agreed, the maquette was then scaled up to a full-size sculpture of the final design. Martin Rezard of Millennium FX was the lead sculptor, creating a complete, full-sized Cyberman in clay over the frame of a 'human' dummy body.

Moulds were taken from this and used to create the various components of the full Cyberman costume. The final pieces were made from fibreglass. To make the Cybermen seem as if they had actually been constructed from metal, powdered aluminium was added to the final layer of fibreglass. Each separate component of the suit was then hand polished until it gleamed. The only exception was the hands, which were gloves made of soft, silver-tinted silicone.

Ten Cybermen costumes were made, including the slightly different design of the Cyber Controller, with its distinctive visible brain and the pipes connecting it into its life-support systems. The head of each Cyberman was constructed from nine different sections. The rest of the body was built from over 40 component pieces.

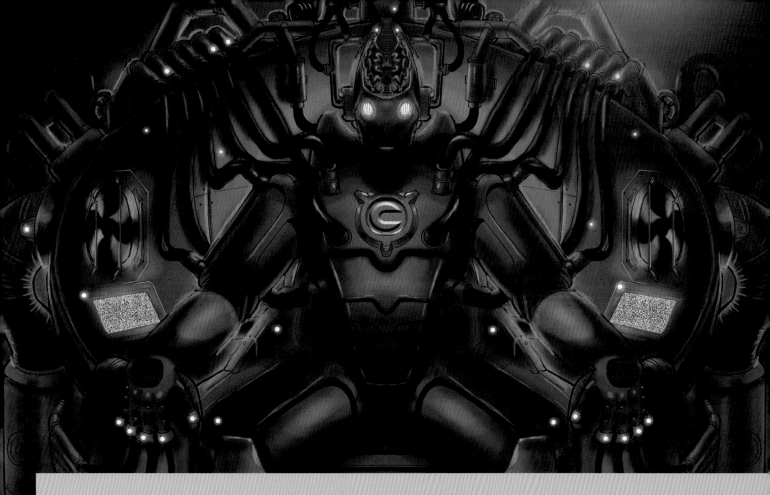

RISE OF THE CYBERMEN

THE CYBERMEN enter the room. Tall, steel giants. Impassive metal faces. Hints of Art
Deco in their design. Cyberman after Cyberman after Cyberman, as many as possible.

BEHIND THE SCENES

The Cybermen costumes created for *Rise of the Cybermen* were reused until 2013, with slight modifications such as the removal of the stylised 'C' Cybus logo on the chest. The Cyber Controller's head, painted partly black, with its visible brain became the head of the Cyberleader in *The Next Doctor*, and the Cybermen – and Cyberleaders – appeared in short sequences in several stories including *The Pandorica Opens* and *A Good Man Goes to War*.

But for *Nightmare in Silver*, a new design of Cyberman was introduced. The previous versions were seen as exhibits and relics. But the new Cybermen were to be sleeker and faster, slimmer and more elegant, albeit while retaining the expressionless face and 'dead' eyes…

Again, there was an elaborate design phase with several concept artists working on different potential designs. Photoshop paintings, 3-D models and clay maquettes were all produced as the possible designs were whittled down to the final version – a design created by Neill Gorton.

This time, only the left-hand side of the Cyberman was sculpted full-size in clay. A digital scan was then made of each element and these were machined-out in rigid foam, with the 'missing' right hand side created from reversed left-hand side elements. Fine detailing was added by hand and moulds were taken from these components. The result was a tight-fitting, sleek costume with sharp lines.

The previous Cyber suits had been constructed from fibreglass, making them inflexible, heavy, and noisy when they moved. This time the suits were made from flexible polyurethane rubber. The resulting suits were lighter and more comfortable for the actors to wear. This was not a new material, but a new paint process (developed in the USA for use in the car industry) meant that it could be made to look suitably metallic, which had not been possible back in 2006. •

NIGHTMARE IN SILVER

The New Model Cyberman is HUGE and Artie is little and he hasn't turned to see it, and now two silver hands come down to his shoulders, and he starts to scream, but a silver hand covers his mouth.

The Daleks

DALEKS – THE VERY name sends a shiver down the spine. No one who lived through those dark days in 2009 in areas occupied by the Daleks can ever forget. The rest of the world may have moved on, but millions of people are scarred for ever by the experience of Dalek occupation.

But apart from terrifying memories, apart from knowing they exist, apart from the ever-present fear that they might return, what do we actually know about the Daleks?

While everyone is aware of what a Dalek looks like, there is still a misconception among many that the creatures are robots. Nothing could be further from the truth. The armoured shell is merely a container, a life-support system and transport for the grotesque, mutated creature that exists inside. A creature of pure hate, desperate to dominate or exterminate all other life forms. Technologically advanced, the Daleks are the ultimate conquerors, creatures that will not rest until they have brought the entire universe under their control.

Although their hierarchy seems to have changed over the centuries, the Daleks appear to be ruled by an Emperor who acts as head of a Supreme Council – presided over in turn by the Dalek Supreme. This is replaced later by a Dalek Parliament, answering to a Prime Minister, but retaining the Dalek Supreme as a second in command.

The design of the Daleks' outer casing has altered as their technology advances, but the basic form has remained the same. With the ability to levitate and travel through space, the casing has a built-in eye-stalk, manipulator arm (which can be replaced with other tools and appliances), and weaponry. Powered by a form of static electricity, Daleks can survive for centuries without access to an energy source and then revive when power is restored.

Both UNIT and Torchwood were in the front line of the 2009 invasion. From the files and records of both organisations we can be certain that they were aware of the Daleks and the threat they posed long before that. There are records of Dalek incursions dating back to ancient Egypt. And that's just the incidents we know about.

There are many legends and myths about the origins of the Daleks. Some say they were created as part of a super-evolutionary experiment by the alien Halldons on the planet Ameron. Others that they are the mutated survivors of a nuclear war that managed to crawl inside war machines created by a scientist named Yarvelling. The ancient records of the Thal people suggest that the Daleks were once called Dals.

As with many myths, there is a grain of truth in all these stories. The Daleks are indeed the result of genetic experimentation, but not by the Halldons. They did emerge as survivors from a terrible nuclear war, but the Dalek casings they inhabit were always designed to act as life-support systems for these mutant creatures. And the Daleks did have another name before they became the horrific creatures we know today, but they were not called Dals. They were Kaleds.

Thousands – perhaps millions – of years ago, there was war on the planet Skaro. For a thousand years the Kaleds and the Thals fought against each other. Gradually, it became a war of attrition, both races reduced to a single great city facing each other across a desolate wasteland. Nuclear, chemical and biological weapons had already started a cycle of mutation among both populations, and each banished the resultant 'Mutos' out into the wastelands to fend for themselves.

But the Kaleds' chief scientist, Davros, realised that the cycle of mutation was irreversible. He experimented on living tissue to determine the final form his race would take. Then he worked t devise a travel machine and life-support syste that would ensure the survival of his race.

Davros — creator of the Daleks

'Mutos' were banished to the wastelands of Skaro

Crippled, and confined to a mechanised wheelchair and life-support system himself, Davros had some affinity with the mutations. He fashioned a travel machine partly in his own image, basing it on his own systems. Determined that his race should survive, he introduced genetic changes into the mutated creatures he had developed. He removed perceived weaknesses, like pity and conscience. He created a monster that hated every other life form and was desperate to survive by conquering all other life in the universe. Only when the creatures were the dominant race in the universe, Davros reasoned, would there be peace... Having altered the Kaled form so drastically, Davros gave his creature a new name – the Dalek

When his own people, the Kaleds, discovered what he was doing, they tried to stop him. But to prevent this, Davros betrayed them to the Thals and the Kaleds were destroyed. Davros then turned his new Daleks loose on the Thals – and all but destroyed them.

Davros planned for every contingency, except one. Genetically programmed to regard all other life forms as inferior, the Daleks turned on their creator and shot him down. But, as it later transpired, like his Daleks, Davros was not easily destroyed...

FROM: COUNCILLOR MOGRAN

TO: SCIENCE COMMITTEE, WAR CABINET

As you are aware I met today with Davros at the Bunker to inform him of the committee's decision to mount an investigation into his work following the unsettling information received from agents Doctor and Sullivan.

I was surprised at his equanimity and the degree of cooperation afforded by both Davros and Security Commander Nyder. While stressing that all research and development must cease for the duration of the investigation, I have agreed to a 12-hour period of grace as Davros assures me it will take that long to shut down certain equipment. That will give myself and the Tribunal time to assemble and gather information.

I should add that, despite the apparent cooperation, paperwork we have obtained, together with other testimonies supports the contention that the Dalek Project is not only diverting vital resources from more essential areas but is also a danger to ourselves, to our war effort, and to the continued survival of the Kaled race as we know it.

I shall of course keep you updated on our progress and will forward preliminary findings within the week.

Mogran

A confidential memo sent by Councillor Mogran, the senior official given the task of evaluating Davros's work. The Kaled race was almost completely destroyed in a Thal attack orchestrated by Davros before Mogran's tribunal could start work.

Photographs of the Mark 3 Travel Machine leaked to Mogran's tribunal. By this stage Davros had already decided to call it the Dalek.

The Thals documented their history, recording events going back almost half a million years. Earlier sections are distilled from oral accounts and traditions and are therefore less accurate.

WHEN THE MIGHTY EXPLOSION DESTROYED THE WHOLE OF THE CONTINENT OF DARREN, ONLY A FEW THALS SURVIVED. THEY MADE THEIR WAY TO THE CONTINENT OF DAVIUS, SPENDING MANY YEARS ON THE JOURNEY, CROSSING THE ISLANDS OF MOVING MOUNTAINS, AND JOURNEYING ALMOST TO THE RIVER OF WHIRLING WATERS. HERE THEY BECAME FARMERS AND SURVIVED FOR MANY GENERATIONS ON THE CROPS THEY SOWED.

BUT AS THE RAINS STOPPED AND THE HARVESTS FAILED, THE THAL PEOPLE WERE FORCED TO MOVE AGAIN, IN SEARCH OF BETTER SOIL AND MORE CLEMENT WEATHER. THEY TRAVELLED TO DALAZAR, IN THE SHADOW OF THE DRAMMANKIN MOUNTAINS AND CLOSE TO THE LAKE OF MUTATIONS.

THE LAKE WAS FILLED WITH THE MOST TERRIBLE CREATURES, AND MANY THALS WERE LOST AS THEY SEARCHED FOR FOOD OR TRIED TO FISH THE DEEP WATERS. FINALLY, THE THAL PEOPLE, LED BY TEMMOSUS THE WISE, REACHED THE GREAT PETRIFIED JUNGLE. THE TREES WERE AS BRITTLE STONE, UNSTIRRED BY THE WIND AND PRESERVED BY THE SAME TERRIBLE WAR THAT ALMOST DESTROYED THE PLANET OF SKARO.

AND HERE THE THALS FOUND THE DALEK CITY, SPREAD OUT BEFORE THEM ON THE DESERT SAND AND REACHING DOWN INTO THE DEPTHS OF THE EARTH.

45

"There came at that time creatures from the beyond the skies. Fashioned from metal, moving without the need for legs, they spat fire from their strange fingers and spoke with a mighty voice from out of the air. Khephren, the great architect, determined that these creatures and the people with them had come to steal away the treasures of our beloved Pharaoh that were to accompany him on his journey to the Afterlife.

So mighty Hyksos, bravest of warriors, led his soldiers in battle against the sky-creatures. But even Hyksos and his captain Tuthmos could not kills the beasts. And much blood was spilled on the desert sand before the creatures departed once more for their home beyond the sky."

An inscription recently discovered within a secret passage of the Great Pyramid. The translation is by Tobias St John, who dates the original hieroglyphics to approximately 2600 BC.

The only part of Theodore Maxtible's journal to have survived the explosion that destroyed his house, just outside Canterbury, in June 1866.

2 May 1866

Waterfield agrees that the reasoning seems sound. To summarize: A mirror reflects an image. So one may be standing in one place, and yet appear to be standing 50 feet away or more. We should, therefore be able to refine the image in the mirror, and then project it. In so doing we shall be able to travel through time just as the reflected image appears to travel through space!

And so we have built the cabinet, and filled it with 144 separate mirrors (sheets of polished metal). As like repels like, we should be able to subject the polished metal sheets to an electrical charge and then repel the image in the mirror wherever we direct.

16 May 1866

Both positive and negative charges have now failed. The only alternative we have not yet tried is static electricity. Tonight, we shall attempt our final experiment.

17 May 1866

Success! But at such a cost. Midway through our experiment, creatures appeared, coalescing from the reflected images and bursting forth from the cabinet. These 'Daleks' are from another place, another time.

They have taken Waterfield's daughter Victoria and hold her hostage. But I believe I can reason with them, and forge a bargain of necessity. I know what they want, and we can help them achieve their ends. By means of the cabinet, Waterfield will travel forward to the time they specify, and entrap the Doctor and Macrimmon

Photographs retrieved from the ruins of the Dalek City on Skaro by a Thal expedition. The Dalek Civil War has been linked to the Maxtible Incident of 1866.

Originally named Amazon, the ship was renamed Mary Celeste in 1869. It is a name that resonates through maritime history. The ship was discovered about 600 miles west of Portugal, drifting and abandoned, by the Dei Gratia on 4 December 1872. Many theories have since been put forward to explain what happened to the ten people on board, including Captain Briggs's wife and infant daughter.

While there had been some bad weather reported in the Atlantic, it seems unlikely this was to blame for the disappearance of the crew. Other theories include underwater earthquakes, alcohol fumes or drunkenness, a freak water spout, mutiny, piracy, attack by some marine creature possibly a giant squid… Perhaps most outlandish of all is the suggestion that the crew was frightened overboard, or possibly even abducted, by space aliens.

The unfortunate ship herself, having already been salvaged after running aground in 1867, was deliberately destroyed in 1885 in an attempted insurance fraud.

Although Miles Pettifer scoffs at the notion in his *Mysteries of the High Seas*, there is some evidence to suggest that the crew of *Mary Celeste* was attacked by Daleks.

HOOVERVILLE PIG MAN

The scores of homeless families camped out on the Great Lawn in Central Park may have more to worry about than poverty and starvation. For several weeks there have been reports of a 'Pig Man' being seen in the area. Descriptions remain vague, but the man is said to be of average height, with a stocky build, and the face of a pig.

The Pig Man has already been linked to several disappearances, the latest being a stagehand who went missing from the Laurenzi theatre. Is this beast now preying on the disaffected and impoverished residents of the Central Park Hooverville?

To add to the suffering, there have been reports that a fire at the settlement last night resulted in the death of a prominent resident known as Solomon. There are unconfirmed reports that local businessman and philanthropist Mr Diagoras was also killed in the fire.

From a 1930 edition of the *New York Gazette*. Diagoras has been linked to possible Dalek activity at the near-complete Empire State Building.

Blueprints for Professor Edwin Bracewell's Ironside war machine.

FRONT ELEVATION

SIDE ELEVATION

New York Revue

STARING BARKER'S BELLES

AT THE LAURENZI THEATRE

The Laurenzi theatre was another site linked to Dalek activity in 1930s New York.

WRITTEN AND
DIRECTED BY
MR FRANZ WHEELER

INTRODUCING
BARKER'S
BELLES

LORETTA FORBS
TALLULAH FRANC
LOIS SHAW
MITZY SEGAL
TABITHA LEE
SUSI BLANCO
FRANKIE MANELLI
ELLA LANE

A memo from noted physicist and scientific military intelligence expert Reginald Jones to Churchill and his Chief of Staff. The Ironside machines that Edwin Bracewell claimed to have invented were actually Daleks. Their exact mission has not been established.

FOR EYES OF PRIME MINISTER
AND GENERAL ISMAY ONLY

I have today met with Bracewell to see a demonstration of his new Ironside machine, and can confirm it to be every bit as impressive as we have been led to believe. Bracewell insists there is still development to be done, but I see no reason why we should not begin immediate mass production of the machines.

As you requested, I have also asked Bracewell to instigate a feasibility study into the use of the attached weaponry as a 'stand-alone' firearm. My understanding is that the size and weight of the power supply, currently housed within the main armoured casing, may preclude this.

R V Jones

Following the Shoreditch Incident it seems prudent to establish the ad hoc Intrusion CounterMeasures Group (ICMG) on a more permanent basis. As per my previous report, I recommend that the current personnel be seconded immediately as full-time members of the group. (I note your comment that Professor Jensen may be resistant to this idea. If necessary we can invoke the Peacetime Emergency Powers Act.)

The ICMG would therefore comprise:

Group Captain Ian Gilmore – in overall charge of all military aspects of ICMG's work. He will be responsible for recruiting his own support platoon from RAF / Army units.

Professor Rachel Jensen – in overall charge of all scientific aspects of ICMG's work.

Dr Allison Williams – to assist Professor Jensen.

As there is likely to be some understandable tension between Gilmore and Jensen, and to ensure political oversight, I further propose that ICMG be under the overall control of a senior civil servant. Someone of the calibre and perspicacity of Sir Toby Kinsella would be ideal.

Part of a report into the 1963 Shoreditch Incident, during which two distinct Dalek factions engaged in a pitched battle on the streets of London. ICMG seems to have been a forerunner of UNIT.

With the postponement of the International Peace Conference in London after the mysterious death of the Chinese Delegate and an attack on the American Senator Alcott, the international situation deteriorated rapidly. This urgent message was sent to all national UNIT HQs ahead of the reconvened conference where a peace deal was finally brokered by Sir Reginald Styles.

Emergency Message to all UNIT National HQs. FLASH URGENT.

Latest official report: The international situation is growing steadily worse. War now seems inevitable. As yet, there is no further news of Sir Reginald Styles, who has flown to Peking in a last minute attempt to persuade the Chinese to reconsider their withdrawal.

Observation satellites report troops massing along the Russian-Chinese frontier. In South America and southern Asia, reports say fighting has already broken out in many regions.

All UNIT personnel are hereby placed on maximum alert.

When Colonel Archer's bomb disposal team failed to report in on schedule, I despatched Captain Winterton to investigate. It never crossed my mind that there was any problem. The making safe of unexploded ordnance is of course always a risky business. But the strange canisters Archer was investigating at an abandoned warehouse close to the Thames did not seem to be dangerous.

It was therefore with some surprise that I received Winterton's urgent message requesting that I join him at the site. On arrival, I found Winterton, and his troops, in sombre mood. On entering the warehouse, they had found the bodies of Colonel Archer and his men, including Sergeant Calder. One of the privates had grotesque claw marks on his neck, but there was no sign of how the others had met their deaths.

On an upper floor, we found Professor Laird – also dead, but unlike the others she had been shot in the back. A subsequent forensic analysis revealed that it was actually Archer's sidearm that had fired the fatal bullet.

But the most bizarre discoveries were yet to come. In the area where the canisters had been discovered, Winterton's men had found several strange machines. Each standing over five feet high, with studded lower sections and various appendages. They seemed inert, and several had been destroyed by explosives – revealing grotesque, organic innards. Following standard procedure for reporting anything this extraordinary, I immediately gave orders for C19 to be informed.

Within minutes, Sir John Sudbury himself made contact with us. We were ordered to leave the warehouse immediately and take no further action. My protestations were overruled, and we were obliged to leave behind the bodies of our fallen comrades in arms. Although officially I have no knowledge of what happened to the site, as we were leaving several vehicles with UNIT insignia arrived at the scene.

ALTRON REF. 00.1 00.2 00.3 00.4

METALTRON
FRONT ELEVATION

REF. 00.1
00.2

00.3

00.4

00.5
00.6

00.7
00.9

00.10

00.11

00.12

00.13
00.14

These plans of the 'Metaltron' may have been drawn up by by Van Statten's adviser Adam Mitchell.

RADIOGRAPHER PRESENT: M.SAVAGE 098.798904
SUBJECT: METALTRON
SCAN: SIDE ELEVATION

00.009.893-028949203

RADIOGRAPHER PRESENT: M.SAVAGE 098.798904
SUBJECT: METALTRON
SCAN: FRONT ELEVATION

00.009.893-028949203

STRICTLY PRIVATE AND CONFIDENTIAL

To: Henry Van Statten
For sale by Private Auction
A Once-In-A-Lifetime Chance
to Own an Alien Creature

"The Metaltron"

Provenance: The Metaltron's provenance is well known among serious private collectors. It fell to Earth in 1961 on the Ascension Islands. The creature burned in its crater for three days before anyone could get to it. All that time it was screaming – insane. It has remained in various private collections ever since.

**REPLY BY EMAIL TO REGISTER YOUR INTEREST.
$1,000,000 FEE TO ACCESS THE AUCTION WEBSITE**

Billionaire Henry Van Statten bought his 'Metaltron' — once believed to be the last Dalek survivor of the Great Time War — at a private auction organised over the internet.

remembered for the invasion of the Cybermen, it should not be forgotten that a group of Daleks also arrived in the Torchwood Tower. According to some sources, they were housed inside a mysterious metal sphere, which Torchwood scientists had been studying for some time and which may have played some part in the weakening of the fabric between universes.

The testimonies of survivors from within the Tower suggest that the Daleks regarded the invading Cybermen as little more than 'pests', and dealt with them accordingly. The Cybermen seem to have been vulnerable to Dalek firepower, and unable to fight back effectively.

Some survivors were also able to provide information about the Daleks themselves. It seems there were four of them in all – led by a Dalek with a black casing, as opposed to the more usual bronze armour. These Daleks formed the Cult of Skaro, a group established by the Dalek Emperor to dare to *imagine*. They tried to think like their enemies, to understand their thought processes and determine their strategies. To this end these Daleks even took names. Led by Dalek Sec, the other Daleks were Thay, Caan, and Jast (some reports mention a Dalek Rabe, but this seems to be an error).

What happened to this Cult of Skaro is not known. They – and many more Daleks who emerged from a capsule above London during the Battle of Canary Wharf – disappeared along with the invading Cybermen. Perhaps the Cult of Skaro was actively involved in the subsequent invasion of Earth. Was this first incursion a fact-finding mission preparatory to the main invasion? Whatever the case, the Daleks to return in force, and this time the area they attacked was New York in ruins, devastated by the

The striking frontispiece to Peterson's book depicts the horrific events of the Dalek invasion.

The 2009 invasion provides the most comprehensive evidence of Dalek activity. Thousands of photographs were taken during the invasion, some of them on board the Dalek Crucible Ship.

Images from the security systems in Torchwood Tower, retrieved after the Battle of Canary Wharf.

Some commentators have asked why the Osterhagen Keys were not used during the Dalek invasion. Whatever the reason, we can remain thankful that Osterhagen was not activated.

The Osterhagen System is potentially far more effective than the Z-Bomb system it replaces. It also includes a far greater level of security and safeguard against accidental or terrorist use.

Whereas the Z-Bomb was under the control of the individual commanders, and any one Z-Bomb could be detonated in isolation with catastrophic effect, Osterhagen requires the simultaneous use of 3 of the 5 keys. Once 3 keys are operated in unison, the chain of 25 nuclear warheads placed around the globe will detonate ensuring absolute and total destruction of the planet.

There is obviously a subjective, human element to any solution of this magnitude and severity. But the intention is to activate Osterhagen if – and only if – the suffering of the human race as a whole demands that we deny our planet to an extraterrestrial aggressor.

I was excited by the capsule – 200 years in a mercury swamp and not even tarnished. But this!

I've not told anyone I have been inside the capsule. Except Janley and Resno of course. We have to be careful, especially with an Examiner due any time. But I have removed one of the three robotic machines. Perhaps if we can restore power to it... Who knows, the capsule has withstood rain, damp, heat, mercury – could the robot still be viable?

Notes taken by the scientist Lesterson, who was responsible for reviving a squad of Daleks on the colony planet Vulcan. Despite Lesterson removing their weapons, the Daleks almost wiped out the entire colony.

The Examiner is a strange man. He is trying to tell us the robots – 'Daleks' he calls them – are dangerous. Thank goodness he doesn't know I have already removed one. I wonder how it operates – a simple positronic brain, I shouldn't wonder...

Images from the Vulcan Colony security systems.

We attached power to the Dalek, and at a level of 3.24 it reacted! I designated the three attachments on the front of the machine with numbers. One was a manipulator of some sort. Three incorporated a lens – so an optical device. Number 2 attachment I had no idea about until now. But it is a weapon! When the power reached 4.68, the machine shot Resno. Janley tells me he is recovering. I pray she is right.

A brief account of the Dalek invasion of Earth in 2157, taken from *The Rise and Fall of Earth's Empire* by Layten Halkovich.

The meteorite storm was the first indication that anything was wrong. Even then, it was assumed to be a natural phenomenon – some sort of cosmic storm. When the meteorites eventually stopped, everything settled down again. But then people began to die of a new and unidentified plague.

In the worldwide epidemic that followed, whole continents of people were wiped out. Asia, Africa, and South America were all depopulated. In just a few weeks, Earth had a stench of death about it. By the time a cure was found, based on the rare chemical parrinium, it was too late. As well as killing 90 per cent of Earth's population, the plague split the world into small communities, too far apart to combine and fight. Too small individually to stand any chance against invasion.

Just six months after the meteorite fall, the Daleks landed in force. Some of the surviving cities were razed to the ground, while others were simply occupied. Anyone who resisted was exterminated. Some people were captured and were turned into Robomen, mindless slaves of the Daleks with no will of their own, compelled to obey without question. They caught other human beings and many of them were shipped to the vast mining areas. Almost no one escaped. The Robomen saw to that.

Probe Survey #7768/R

Data returned by an automated probe sent to check on the status of potential colony planets.

PLANET: ++ Mechanus ++

DESIGNATED: ++ Colony Potential ++

STATUS: ++ Mechonoid Robots despatched prior to Interplanetary Wars ++

MISSION: ++ Clear landing sites for first immigrants ++

SURVEY RESULTS: ++ Evidence of pathway and landing site clearance through dense jungle areas. ++ Debris and rubble detected across a wide area at centre of clearance sites ++ Uploading image data from inactive Mechonoid storage systems ++

CONCLUSION: ++ Construction of city to receive first colonists completed ++ City then destroyed in catastrophe ++ Analysis of debris suggests use of heat weaponry and explosives ++ Mechanus may be in a conflict zone and should be redesignated Category C Status ++

The image data recovered by the remote probe, once reassembled, shows what happened on Mechanus.

Images brought back by survivors of the Thal mission to the planet Spiridon in the Ninth System.

Fragment of a Dalek space signal sent from the planet Spiridon to Dalek Supreme Command, intercepted by the Thals.

ANTI-REFLECTING LIGHT WAVE DERIVED FROM SPIRIDON LIFE FORMS ENABLES DALEKS TO ACHIEVE TOTAL INVISIBILITY FOR PERIODS IN EXCESS OF TWO WORK CYCLES. RAY EXHAUSTION AND LIGHT WAVE SICKNESS DUE TO POWER DEPLETION REMAIN A PROBLEM.

TWO THAL PRISONERS TO BE TRANSFERRED TO SECTION 3 AFTER INTERROGATION FOR EXPERIMENTS OF LIGHT RAY EMISSIONS ON LIVING TISSUE. ONLY TWO THALS ARE NOW BELIEVED TO REMAIN AT LIBERTY. THEIR CAPTURE IS IMMINENT.

FORCE ASSEMBLED ON SPIRIDON NOW COMPLETE. 10,000 DALEK UNITS PREPARED AND READY FOR THE INVASION OF ALL SOLAR PLANETS.

Marc Cory, Special Security Service, reporting from the planet Kembel. The Daleks are planning the complete destruction of our galaxy. Together with the powers of the outer galaxies, a war force is being assembled. If our galaxy is to be saved, whoever receives this message must relay this information to Earth immediately. It is vital that defence mechanisms are put into operation at once. Message ends.

Mavic Chen, Guardian of the Solar System, who betrayed humanity to the Daleks.

But this victory has not been without cost. We may celebrate the death of traitors like Mavic Chen, but the loss of loyal, brave agents is always a tragedy.

Marc Cory — More than anyone, it was Cory who first realised the extent of the threat and the possibility of Dalek operations on Kembel. He was not afraid to take immediate action without thought for his own safety. His sacrifice is a lesson to us all.

Kert Gantry — Seconded to the 50A mission and missing in action on Kembel, presumed exterminated. He will be missed by everyone in SSS and especially his fellow agents in CCE Division.

Bret Vyon — One of our most accomplished and courageous agents. Betrayed by his own Commander in Chief and denounced as a traitor, he deserved better from us all.

Sara Kingdom — Tricked into assassinating her own brother, Bret Vyon, no one can doubt the bravery and dedication of Agent Kingdom. She sacrificed her life on Kembel to save the Solar System and Aligned Worlds from the Daleks and their allies. We extend our deepest sympathy and everlasting gratitude to her friends and family, in particular her husband Agent Tarrant Kingdom.

Part of a bulletin sent out to all Special Security Service personnel following events on Kembel.

Delegates of the Dalek Alliance assembled on Kembel. They have been provisionally identified as (from left to right): Beaus, Celation, Trantis, Malpha, Gearon, Sentreal. The Daleks betrayed and exterminated them all — including Mavic Chen.

44

From *The Lives and Deaths of Davros,* by noted Dalek historian Lorraine Baynes.

FROM A MODERN PERSPECTIVE it seems unlikely that the Daleks were ever truly threatened by the Movellans. But there is no doubt that the robot race caused considerable disruption to Dalek operations, and the Daleks resorted to drastic measures to gain supremacy.

With their battle systems locked in a digital stalemate, the Daleks despatched a small task force to their original home planet Skaro. Long since abandoned, Skaro was a desolate wasteland. But using slave labour, the Daleks burrowed into the buried remains of the original bunker where Davros first created them. And this was what they sought to retrieve — the remains of Davros himself.

Revived from dormancy, his life preserved by the systems that had kept him alive through the latter stages of the Thal-Kaled war, Davros attempted to take control of the Dalek race. But the Daleks' slaves revolted, destroying a force of Movellans and capturing Davros. He was sent to Earth where he was put on trial for his war crimes...

The Dalek Wars involved billions of troops and thousands of warships deployed across vast areas of space to stop the spread of the Dalek Empire. At the height of the Wars, *Aristotle* claimed a rare victory against superior Dalek forces.

Deep Space Cruiser Aristotle reports significant Dalek activity in the Ryzak solar system. Aristotle remains on station, shielded and hidden within an asteroid belt. Colonel Morgan Blue commanding. Further information to follow...

Images downloaded from the Prison Station's security feed before it was destroyed.

1600-PR   14

→13A   14   →14A

From *The Lives and Deaths of Davros*, by noted Dalek historian Lorraine Baynes.

59

DESPITE HIS CRIMES, AND evident lack of contrition, capital punishment was ruled out on moral grounds. Instead, Davros was cryogenically frozen and imprisoned in a secure space station, its location a closely-guarded secret.

It took the Daleks almost ninety years to locate Davros. But they are nothing if not persistent. After such a long period, the prison station had fallen into disrepair and the crew into complacency. Although the Officer of the Watch was astute enough to launch fighters to investigate when a ship approached, the pilots and their craft were no match for the Dalek Deep Space Cruiser that attacked. The station was soon overrun, and Davros revived.

With rising Dalek casualties from the virus developed by the Movellans, the Dalek Supreme demanded that Davros develop a cure. But again he attempted to take control, turning the Daleks against each other. He even released samples of the Movellan virus to destroy Daleks that were not willing to submit to his command.

The Dalek ship was eventually destroyed when the prison station's self-destruct systems were activated, presumably by a surviving member of the crew. Although the Dalek Supreme and the main task force were destroyed, the results of Davros's analysis of the virus had already been transmitted to Dalek Supreme Command, and a cure was rapidly synthesized.

DEAR AND NOT YET DEPARTED
Need a break? Ready to put your life on hold?
Waiting for a cure for an inconvenient or fatal ailment?
Then you need:

TRANQUIL REPOSE

Situated on the picturesque planet Necros, Tranquil Repose offers a unique service. Personally supervised by the Great Healer himself, your body can be cryogenically stabilised until you are ready to return to life.

Your resting consciousness will be constantly updated concerning social, cultural and technological developments. You won't wake up feeling the universe has left you behind.

AND FOR JUST A SMALL EXTRA COST, YOU MAY PURCHASE OUR PERSONALISED COMMUNICATION SERVICE.

The galactically famous DJ says: "If you're missing your resting one and want to tell them just how much, then why not call on me? I am the messenger who connects your heart with their heart."

Tranquil Repose – for anyone who is tired of life, but hasn't finished living.

Davros escaped to the planet Necros where he became 'the Great Healer'.

Images from inside Tranquil Repose, recovered from security cameras after the facility was destroyed.

OFFICIAL ANN...

IT IS WITH THE DEEPEST REGRET THAT THE GRAND ORDER OF OBERON ANNOUNCES THE DEATH OF ONE OF ITS SENIOR KNIGHTS. ALTHOUGH ORCINI HAD BEEN TEMPORARILY EXCOMMUNICATED FROM THE ORDER, HE CONTINUED TO ABIDE BY OUR RULES.

Together with his Squire, Bostock, Orcini undertook the noble assignment of the assassination of Davros, creator of the dreaded Daleks. Although he did not succeed, he did manage to destroy a vast army of Daleks created by Davros. 'Not since the days of Abslom Daak,' said Supreme Master Talcadian, 'have so many owed so much to one of our esteemed Order.'

According to tradition, Orcini's name has been added to the Roll of the Glorious Dead in the Great Hall of Oberon.

A statement put out by the Grand Order of Oberon.

<CONT. FROM PREVIOUS

Space Traffic Control's report into the loss of the starliner *Alaska*.

No signal has since been received, and so we have to conclude that, tragically, the starliner *Alaska* was lost with all souls. Navi-Computer projections from her last known position and course, taking gravitational variations into account, suggest that the ship may have fallen into the attraction field of the planet now known to be the Dalek Asylum.

If that is the case, then an MSC rescue mission is out of the question. Even if, by some miracle, anyone survived the impact, they will soon have succumbed to the latent nano-field which surrounds the planet, and undergone Dalek conversion. Not that this will save them from being hunted down and exterminated by the insane creatures that even the Daleks themselves belief to be a danger to all of creation.

Our recommendation therefore is that *Alaska* is listed as 'Missing' with no further details made public.

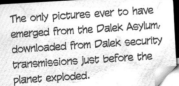

The only pictures ever to have emerged from the Dalek Asylum, downloaded from Dalek security transmissions just before the planet exploded.

From *A History of the Dalek Wars*, the definitive account of the various conflicts, by Hagan Garsonomous.

THE THIRD DALEK WAR

The last action of the Third Dalek War was the Exxilon Gambit. Again, the Daleks employed a form of germ warfare, infecting the Outer Worlds with a plague that killed the colonists in their thousands. With millions more about to perish, the humans desperately sought for – and found – a cure. The rare chemical parrinium could halt the disease, curing those already sick and providing immunity to others. A source of parrinium was located on the remote planet of Exxilon.

But once again, the Daleks had planned for this contingency. A squad was despatched by supra-lightspeed saucer to secure the parrinium so that it could be used to blackmail the Space Powers into surrendering.

Conditions on Exxilon were harsh, and the Daleks found themselves deprived of their usual weaponry and forced into an uneasy alliance with the Marine Space Corps group charged with recovering parrinium from the planet. The Daleks claimed they also needed parrinium as inhabitants of their own colonised and occupied worlds were suffering from the same plague. This was of course a ruse, and the Daleks soon reneged on their agreement with the MSC personnel.

It was the bravery of acting commander Dan Galloway, weapons officer on the expedition, that led to the destruction of the Dalek taskforce and the safe acquisition of sufficient parrinium to halt the plague.

ALTHOUGH IT TOOK place largely unnoticed by lesser races, the Great Time War raged throughout all of space and time – a terrible conflict fought between the Daleks and the Time Lords. It was so devastating that the entire war was timelocked to preserve the rest of the universe.

As a result, only small pieces of information have emerged. But from these details it is clear that the war was initially provoked by the Time Lords. Foreseeing a time when the Daleks would have conquered all other races and would reign supreme, the Time Lords sent an agent back in time to the very moment when they were created by Davros. The Time Agent's mission was simple – to avert the creation of the Daleks. He failed.

When they learned of the plan, the Daleks declared war on the Time Lords. It was a war that was believed to have destroyed both races. In the final moments of the war, as Gallifrey's second city Arcadia fell to the Daleks, one renegade Time lord used the ultimate weapon – The Moment – to wipe out both races. Or so it seemed.

But there are rumours that the Time Lords survived in another reality. Certainly the Daleks survived. The Emperor and his crippled command ship fell through time, to create a new Dalek race. The original race also survived, preserving 'pure' Dalek DNA in Progenitor Devices scattered through the Universe.

Even Davros, it seems, escaped – rescued from the jaws of the Nightmare Child by the last survivor of the elite Dalek group known as the Cult of Skaro.

According to some, the last battle of the Great Time War was not the Dalek attack on Arcadia, but the Siege of Trenzalore where a Dalek warship was allegedly destroyed by the Time Lords through a crack in the fabric of reality…

There is no reasoning with a Dalek. It is important to understand that if you ever meet one. The only interest a Dalek is likely to have in you is in your extermination. If the Daleks keep you alive, it is because they want something – information, or more likely cooperation. The Dalek Empire is constantly expanding, and slavery is widely used to provide raw materials and labour.

There is a myth that Daleks are unwieldy and have trouble negotiating rough terrain or stairs. This is not true – the Daleks have anti-gravitational technology that enables them to levitate over obstacles. If they do have a weak point, it is the eye-stalk at the top of the dome, which can be destroyed with a concentrated burst of gunfire or a heavy percussion device such as a bastic bullet.

But it isn't only the Daleks you need to be wary of. Just as they enslave humans, they also use other races as expendable foot soldiers, particularly the ape-like Ogrons. Alliances with the Daleks tend to be one-sided and short-lived, but they are not unknown. Be wary of traitors like Mavic Chen or Theodore Maxtible.

Be wary too of Dalek Agents – converted in some way to obey the Daleks. This conversion may be a simple brain-conditioning, or it may be enhanced by implants. In the Dalek invasion of Earth in the twenty-second century, many humans were converted into Robomen – mindless Dalek slaves. Later developments of Robo-Technology mean that the Daleks can reanimate dead humans, or create duplicate copies. Often there is no way to detect such a Dalek agent until it is too late…

They may have been defeated so far in their attempts to invade Earth and destroy humanity. But they are still out there, planning and waiting. We know there will be long and hard-fought Dalek Wars in the future. We know they will invade Earth again.

The Daleks are coming.

BEHIND THE SCENES

Writer Terry Nation had some experience of writing television science fiction, but he was best known for his work on comedy and light entertainment when he was approached to submit ideas to *Doctor Who*. Busy with other writing commitments, Nation almost turned the offer down. But finding himself unexpectedly out of work, he agreed take on the job. Later he would admit that his intention was simply to 'take the money and fly like a thief.'

Nation's initial storyline went through various revisions – most notably to remove the third race of aliens that originally arrived at the end of the story to confess they had started the Great War and settle the dispute between the Daleks and the Thals.

For his villains, the Daleks, Nation wanted a creature that was different from anything that had been seen before. He was inspired by seeing the Georgian State Dancers – the women in their long skirts seemed to glide across the stage, and this was something Nation took as a starting point. His other objective was that the Dalek should have no recognisably human features. He definitely did not want a creature that was obviously an actor dressed in a suit.

The description in Terry Nation's scripts, however, gave little idea of the tremendous visual impact the Daleks would achieve. But reading it today there is little doubt that Nation is describing what would become a Dalek.

The job of bringing that description to life was given to BBC Designer Raymond Cusick. He based his design around the shape of a man sitting on a chair. To this basic shape he added the sucker arm and gun (originally at different levels) and an eye

at the domed top of the creature.

Building the four Dalek machines was subcontracted to a company called Shawcraft Models. The voices were provided by actors Peter Hawkins, who provided voices for many children's programmes including *The Flower Pot Men* and *Captain Pugwash*, and David Graham who provided voices for various Gerry Anderson series (including Brains and Parker in *Thunderbirds*). Their voices were treated using a ring modulator to achieve the distinctive metallic, grating quality.

Although still instantly recognisable today, the design of the Daleks in their first story was changed slightly over time. For *The Dalek Invasion of Earth*, an enlarged base hid bigger wheels to cope with location filming on the streets of London. A disc was added to the back of the Daleks to explain where their received their power away from the electrified metal floors of their city on Skaro.

For *The Chase*, the power-disc was abandoned and replaced with vertical slats around the midsection of the Dalek. This same design continued throughout the 1960s, and went largely unchanged throughout the classic series of *Doctor Who*.

BEHIND THE SCENES

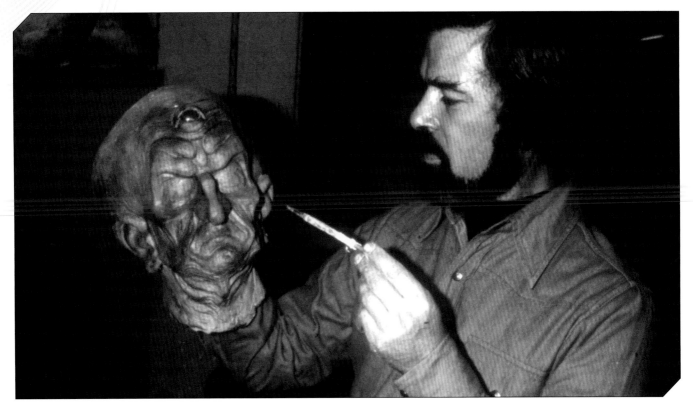

It was the idea of outgoing producer of *Doctor Who* Barry Letts to show the creation of the Daleks in 1975. Enthused by the idea, Terry Nation produced a storyline called *Genesis of Terror*, from which he wrote the scripts for what became *Genesis of the Daleks*. The story changed little during scripting and production. Right from the start, the Daleks were to be the creation of megalomaniac scientist Davros.

'Davros served two roles,' Nation later explained. 'Firstly he was half-man, half-Dalek, a sort of mutated missing link between the two species – the Kaleds as they once were and the Daleks they were to become. Secondly, with the Daleks' slow speech patterns, having a Dalek as a spokesperson for their point of view all through would have been dull, and – worse – would have slowed down the pace of the story. I wanted someone who could think like a Dalek, but talk in a more human fashion.'

The actor chosen to play the crucial role was Michael Wisher. Wisher had appeared several times in *Doctor Who*, and provided Dalek voices for the two previous Dalek stories *Planet of the*

GENESIS OF THE DALEKS
DAVROS IS CONTAINED IN A SPECIALLY-CONSTRUCTED SELF-POWERED WHEELCHAIR. IT HAS SIMILARITIES TO THE BASE OF A DALEK. DAVROS HIMSELF IS A MASTERPIECE OF MECHANICAL ENGINEERING. HIS CHAIR IS A COMPLETE LIFE-SUPPORT SYSTEM FOR THE ANCIENT CREATURE. A THROAT MICROPHONE AND AMPLIFIER CREATE THE VOICE HE NO LONGER HAS. (ITS SOUND IS NOT UNLIKE THE VOICE OF A DALEK.) A MINIATURE H AND L MACHINE KEEPS HIS HEART AND LUNGS FUNCTIONING. A SINGLE LENS WIRED TO HIS FOREHEAD REPLACES HIS SIGHTLESS EYES. LITTLE OF HIS FACE CAN BE SEEN. TUBES AND ELECTRODES ATTACHED TO WHAT DOES SHOW. THE UPPER PART OF HIS BODY IS CONTAINED IN A HARNESS FROM WHICH GREAT COMPLEXES OF WIRES AND TUBES EMERGE.

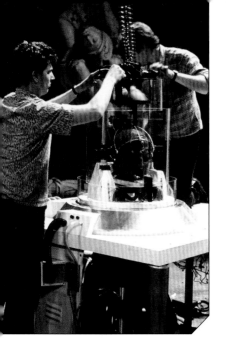

Daleks and *Death to the Daleks*. Wisher rehearsed for his role as Davros wearing a paper bag on his head and sitting in a chair the whole time – which allowed him to perfect his vocal rather than facial performance. In coming up with the distinctive part-Dalek voice, Wisher recalled the tones of the philosopher and mathematician Bertrand Russell. Wisher's voice was fed through a ring modulator to make him sound like a Dalek.

Davros's wheelchair prop was constructed by Visual Effects designer Peter Day and based on a Dalek skirt section. The mask for Davros was created by visual effects sculptor John Friedlander. *Doctor Who*'s producer Philip Hinchcliffe suggested that it should resemble the evil Venusian genius the Mekon, who appeared in the *Dan Dare* comic strips from the *Eagle* comic of the 1950s and 1960s. The mask included a blue bulb which illuminated as Davros' third eye, but this stopped working in some of the recording sessions.

Just as he had with the Daleks themselves, Nation killed off Davros at the end of his first story, not thinking he would need the character to return. But Davros was back in *Destiny of the Daleks*, this time played by David Gooderson. For *Resurrection of the Daleks*, *Revelation of the Daleks*, and *Remembrance of the Daleks* the role was played by Terry Molloy. For the character's return in *The Stolen Earth* in 2008, Julian Bleach played Davros.

THE THIRD DOCTOR

DAY OF THE DALEKS
Written by Louis Marks
Directed by Paul Bernard
First Broadcast: 1–22 January 1970

PLANET OF THE DALEKS
Written by Terry Nation
Directed by David Maloney
First Broadcast: 7 April–12 May 1973

DEATH TO THE DALEKS
Written by Terry Nation
Directed by Michael E. Briant
First Broadcast: 23 February–16 March 1974

THE FOURTH DOCTOR

GENESIS OF THE DALEKS
Written by Terry Nation
Directed by David Maloney
First Broadcast: 8 March–12 April 1975

DESTINY OF THE DALEKS
Written by Terry Nation
Directed by Ken Grieve
First Broadcast: 1–22 September 1979

THE FIFTH DOCTOR

RESURRECTION OF THE DALEKS
Written by Eric Saward
Directed by Matthew Robinson
First Broadcast: 8–15 February 1984

THE SIXTH DOCTOR

REVELATION OF THE DALEKS
Written by Eric Saward
Directed by Graeme Harper
First Broadcast: 23–30 March 1985

BEHIND THE SCENES

The Emperor Dalek first appeared in the Daleks' comic-strip adventures of the 1960s. These epic stories of the Dalek Empire were published in three Dalek annuals – *The Dalek Book*, *The Dalek World*, and *The Dalek Outer Space Book* – and in weekly instalments in *TV Century 21* magazine. The Emperor was shown to be a gold Dalek with an enlarged dome. A similar design, although in white and with no eye stalk, eventually appeared in *Remembrance of the Daleks* in 1988. In this case though, the Emperor's casing turned out to contain Davros.

Remembrance of the Daleks also introduced the Special Weapons Dalek (glimpsed briefly in *Asylum of the Daleks*), a Dalek with no appendages except an enormous gun.

On television, the Emperor Dalek first appeared in the final two episodes of the Second Doctor story *The Evil of the Daleks*. A huge Dalek built into the structure of the Dalek City on Skaro, the Emperor was badly damaged, if not destroyed, when the Doctor infected some Daleks with the 'Human Factor', provoking a Dalek civil war.

It is perhaps intended to be a version of this impressive Emperor Dalek that survived the Great Time War to appear in *The Parting of the Ways*. But whereas his predecessor was a full-sized prop built into the set, the latest Emperor Dalek was a model. Constructed by Mike Tucker and his team at the Model Unit – the same team as constructed the new Dalek prop for its return in *Dalek* – the model was added digitally to scenes where it confronts the Doctor, Rose and Captain Jack. As in *The Evil of the Daleks*, the Emperor was seen to be attended by black-domed imperial guard Daleks.

THE EVIL OF THE DALEKS
WATERFIELD, JAMIE AND THE DOCTOR STAND STOCK STILL STARING
FIRST AT EYE-LEVEL AND THEN THEIR EYES TRAVEL UPWARDS.
NEW ANGLE: NOW TO SHOW THE EMPEROR DALEK, A VAST DALEK
STANDING AT ONE END OF THE ROOM.
WHEN IT SPEAKS ITS VOICE MAKES THE ALIEN BUT MELODIOUS
SOUND OF DALEK VOICES SPEAKING AT ONCE ON DIFFERENT LEVELS.

BEHIND THE SCENES

Almost as soon as the *Doctor Who* production team started planning for the programme's 2005 return to television, they started to plan for the return too of the Daleks. Led by production designer Edward Thomas, the design team set about reinventing the Dalek. Right from the beginning executive producer and lead writer Russell T Davies was determined that the overall shape and instantly recognisable image of the Dalek should remain basically the same. This would be a revision, an improvement on the original rather than a complete reinvention.

The first episode to feature the new design was *Dalek*, written by Robert Shearman. Although the story only features one Dalek, two complete Daleks were actually built – one damaged Dalek, and one in pristine condition.

Working from Matt Savage's original concept drawings and paintings, Mike Tucker and his

team at the Model Unit set about the task of recreating an iconic monster. Using elements from copies of original 1960s Daleks, they updated and improved the Dalek using state-of-the-art materials and technology.

The original Daleks had been operated completely manually by the actor inside. But the team created a radio-controlled version of the Dalek head complete with controllable eye stalk and lights. This allowed actor Barnaby Edwards to concentrate on the other aspects of Dalek movement, although a 'standard' head was also created as an emergency back-up.

As well as a change of colour scheme and extra detailing and design refinements, a 'dog-tag' identification symbol was added under the eye – providing a way for each Dalek to be identified.

BEHIND THE SCENES

Not long after his Daleks returned to *Doctor Who*, Davros was also back. For his reappearance in *The Stolen Earth*, Davros's chair was rebuilt to a similar but updated design by the *Doctor Who* Art Department. A new mask was created by Neill Gorton of Millennium FX. The design was based on the original Davros mask, but updated to make the best use of new materials. The final mask actually consisted of several separate sections, and a glove was made for Davros's prosthetic metal hand. This time it was actor Julian Bleach who brought Davros to eerie life.

The Stolen Earth and *Journey's End*, also saw the Daleks led by the Supreme Dalek. An established part of 'Dalek lore', the Supreme Dalek – sometimes called the Dalek Supreme – was usually depicted as a 'standard' Dalek with a black colour scheme, although the Dalek Supreme that appeared in – and survived – *Planet of the Daleks* was a larger black and gold version.

For *The Stolen Earth* too, a new type of Supreme Dalek was designed. The script specified it should be 'deep metallic red' and the Art Department created several designs before the final look of the Supreme Dalek was agreed. The final version had an extra dome light and bulky side panels that perhaps allowed it to plug into flagship systems.

A more radical redesign appeared in *Victory of the Daleks* and also featured in some subsequent stories. These 'New Paradigm' Daleks were larger, with a more 'moulded' appearance and a colour scheme that reflected the appointed task or rank of

each Dalek. The new Dalek Supreme was white, for example. For its reappearance in *Asylum of the Daleks*, the white Supreme – along with other New Paradigm Daleks that featured – was slightly redesigned to lessen the humped appearance of the back section.

But it is the bronze 'Time War' Daleks that remain truest to the original 1960s design, and which were very much to the fore of the 50th anniversary story *The Day of the Doctor* and the Eleventh Doctor's final adventure *The Time of the Doctor*. ●

The Great Intelligence

IN THE SPRING OF 1935, Edward Travers gave a speech to the Royal Geographical Society in which he stated that 'the body of evidence that has accumulated over the years in undeniable. The Abominable Snowman does exist...' There was uproar. This was not a possibility that the scientific community was willing to countenance. The eminent Professor Walters apparently shouted from the back of the lecture hall: 'If you're as sure as that, my dear Travers, I suggest you go and look for the beast!'

Stung by the criticism and scepticism of his colleagues and peers, Travers accepted Walters' challenge. Together with John Mackay, he went in search of the Abominable Snowman – the fabled Yeti. And in the foothills of the Himalaya Mountains in Tibet, he found it.

The expedition to the Det-Sen area cost Mackay his life. And on his return, Travers' account of his discoveries was treated with as much academic ridicule as his original proposition. Travers asserted that the Yeti did indeed exist. But he had also, he claimed, encountered a group of ersatz Yeti – creatures that looked like the actual Abominable Snowmen of legend, but which turned out to be robots controlled by a nebulous 'Intelligence' that also possessed the Master of Det-Sen monastery. It was only the intervention of a mysterious stranger and his companions that prevented this Great Intelligence from destroying the world...

The story seemed fantastic in the extreme. But Travers brought back evidence from Tibet to lend credibility his story. As well as a variety of advanced componentry, which experts struggled to understand, Travers also brought back a complete inactive Yeti robot. This too baffled the experts, who were unable either to comprehend its design and construction or to get it to work.

Over time, Travers and his story faded from popular perception. Travers himself moved on with his life, but he was never quite able to escape the memories of his past. It was with extreme reluctance that he eventually sold his Yeti robot to the noted collector Julius Silverstein, who gave it pride of place in his private museum.

Can it be simply a coincidence that the unexplained death of Silverstein and the theft of the Yeti exhibit came just a few weeks before the so-called London Menace? The incident was subsequently dismissed as a freak chemical leak which caused the centre of London to be evacuated for several weeks. But at the time there were reports of savage, bear-like creatures being glimpsed through the deadly fog that descended on the capital. Were these the Yeti? Had the Intelligence returned, as Travers later claimed? It is surely not by chance that Travers, assisted by his daughter Anne, was the scientist put in charge of finding a solution to the Menace. The nervous breakdown and subsequent ramblings of Harold Chorley, the only journalist permitted within the affected area, have only added to the speculation.

But do the roots of the London Menace lie even further back than events in Tibet forty years earlier? Do they actually stem from the Great Intelligence Institute of Victorian London, and the legend of the Whisper Men?

Is it just by chance that the name Travers ascribed to the force controlling the robot Yeti in Tibet was the Great Intelligence? The same name as Dr Walter Simeon gave to his academic institute in the early 1890s. Immediately before his sudden death, Simeon himself was linked to the 'Recurring Snowmen' of Christmas 1892, in which snowmen from the previous winter apparently reappeared just as they had been before they melted. Some accounts even maintained that the snowmen could move, terrorising and attacking Londoners... Of course, this was discounted as merely an invention – like the Whisper Men.

Taking their name from an old rhyme, the Whisper Men were invented to keep children in order. Rather like the Bogeyman or Spring-Heeled Jack, they were invoked as monstrous creatures who would deal with naughty children.

Aside from the rhyme, the best-known report of the Whisper Men comes from the story of Clarence DeMarco. DeMarco was convicted of the murder of fourteen women in 1893. His death sentence was unaccountably commuted to life imprisonment, and he went on to live a long if unremarkable life of incarceration, troubled by persistent and chronic insomnia. DeMarco always maintained that he was visited in prison by the Whisper Men the day before he was due to be executed and given a message to pass on to Madame Vastra, the Great Detective (see page 198).

DeMarco described the Whisper Men as figures 'dressed all in black in the manner of undertakers, but with pale featureless faces that might have been made of paper or parchment'. He maintained that these creatures could kill with a whisper, and that they spoke in rhyming couplets – much after the style of the Whisper Men poem...

The UNIT files have little to say about the Whisper Men, although they do include information about the Yeti robots and the Great Intelligence. Although the London Menace predates UNIT by several years, it does seem that the incident was instrumental in UNIT's formation. Indeed, Colonel Lethbridge-Stewart took charge of the military response to the London Menace after the death of Colonel Pemberton. He was subsequently promoted to the rank of brigadier, and became the first commanding officer of the British contingent of UNIT (reporting directly first to Major-General Rutlidge, and later to Major-General Scobie).

But from the UNIT data, there can be no doubt that the Yeti robots did indeed exist, controlled by a disembodied Intelligence, and deployed through London and the London Underground network for reasons that have never been revealed...

The title of Travers' talk – 'On the Abominable Snowman' – had given rise to speculation among members that the speaker would address the subject of myths and legends and how they originate, with particular reference to Eastern cultures and mysticism. However, it was soon apparent that this was not the intention. Indeed, Travers began by enumerating various items of 'evidence' for the existence of the Abominable Snowman, including written eyewitness accounts. The first of these dated back to 1832, from the British Resident Officer in Nepal.

That said, Travers did briefly address the issues of ancient folklore, established over centuries. His description of how the local Tibetans threaten naughty children with warnings of the 'Wild Men' drew comparison from some quarters with the tales of the Whisper Men here in London back in Victorian times.

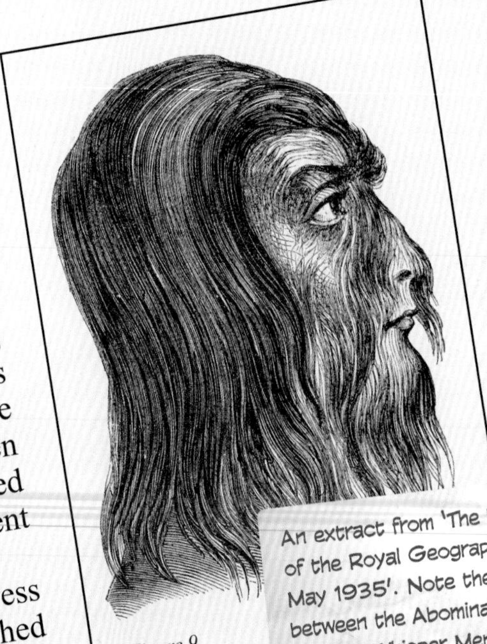

Figure 9

An extract from 'The Proceedings of the Royal Geographical Society, May 1935'. Note the comparison between the Abominable Snowmen and the Whisper Men.

But we now live in a more enlightened age, and so Travers' contention that the body of evidence is now such that we can be sure the Abominable Snowman does indeed exist was met with some scepticism, not to say derision. Professor Walters in particular was most scathing, and suggested in rather forthright terms that Travers might like to find the creature for himself.

Rising to the occasion admirably, Travers accepted this challenge and is understood to be planning an expedition to the Det-Sen region of Tibet, accompanied by noted explorer John Mackay. Whether they will discover the Abominable Snowman is dubious in the extreme. But the Society awaits their return and their account of the expedition with eager anticipation.

Most precious of all the holy relics of Det-Sen is the Holy Ghanta. This prayer bell, so intricate in design, bears the mark of the Great Dragon for it is said the Ghanta was forged in the fire of his breath.

For centuries the monks of Det-Sen have guarded this relic, and the Ghanta has in turn bestowed upon them fortune and safety. But now the Ghanta is lost to us. Now it is gone, removed it is said to a place of safety by one trusted beyond all others by Padmasambhava, the Master of Det-Sen. For our monastery has become a place of danger, attacked by bandits.

Our hopes now lie with the Man of Learning to whom the Ghanta has been given for safekeeping. It is said he will return it to Det-Sen when this danger is past, but as another even greater danger doth beset the monastery.

Part of a seventeenth-century manuscript originating in Det-Sen. This facsimile has been translated into English, and is now sold at the Det-Sen Monastery Gift Shop. Interestingly, Travers' account of his expedition mentions that the Ghanta was returned to Det-Sen while he was there. This would mean it was lost for several hundred years.

Dear Julius

I have today collected together and packaged up the various pieces that I have agreed to pass on to you. As we discussed, I have decided to keep for myself several of the items, including the Yeti miniature figures and three of the metal control spheres together with associated components.

Please find below a list of the items that are now ready for you to collect. I look forward to seeing the various pieces, and in particular the Yeti itself, on display in your museum. I trust that a cheque for the agreed sum will soon be forthcoming.

Yours sincerely

Edward Travers

ITEMS INCLUDED

1 x Yeti / Abominable Snowman automaton (not working)
2 x metal control spheres (inert)
1 x Pyramid structure (damaged)
1 x Det-Sen Monastery monk's habit
2 x Tibetan traditional serving bowls
Several x Ghost Trap ribbons
1 x Rifle (damaged in Yeti attack)

LONDONERS FLEE
MENACE SPREADS

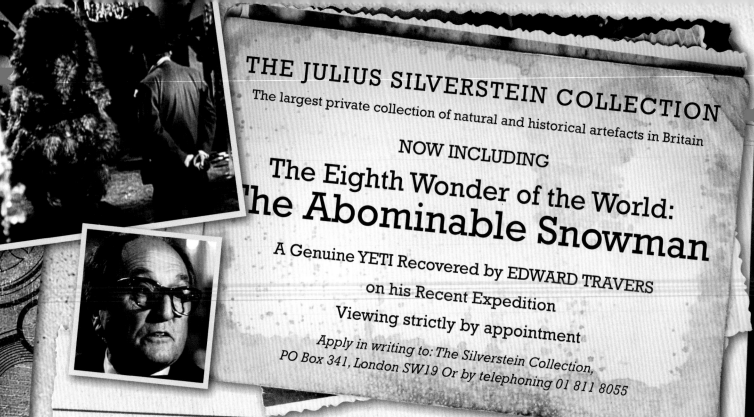

It is now thought that up to 12,000 people died in the Great Smog of 1952. The 'London Menace' did result in some loss of life, especially in military units assigned to the area, and did indeed disperse quickly after several weeks.

Central London has now been evacuated as the thick fog that first appeared on 5 February has continued to spread. The police are warning residents of streets close to the affected area that they should also prepare for evacuation.

The only people known to remain in central London are military and scientific personnel studying the fog with a view to finding a way to dissipate it. While it is not thought to be toxic, the government is warning people to stay away from the mist.

Inevitably, parallels are being drawn with the Great Smog of polluted air that affected London in December 1952 and hastened the introduction of the 1956 Clean Air Act. Approximately 4,000 people died as a result of that smog, with as many as 100,000 suffering ill effects. The Great Smog dispersed quickly after several days when the weather changed. Changes in the weather have not so far affected the current situation

We were, in effect, a base under siege. With fog above ground, and the strange web-like fungus growing through the tunnels of the London Underground tube system, the Goodge Street bunker was a last refuge. All around us, we were acutely conscious, the Yeti watched and waited in the shadows. Where they came from, what they wanted, we could not guess. But Death had taken on frightening, physical form.

The death of Colonel Pemberton in a Yeti attack hit us all badly. There was a palpable change in the atmosphere as morale dipped amongst the troops, although Captain Knight did his best to rally everyone. A personable and talented young officer, I found when interviewing him that he spoke in pure quotes. How trite some of those comments now seem following his own tragic demise…

By this time, we were reduced to a handful of men, plus of course Miss Anne Travers and her father. But it was at this lowest ebb in our fortunes that salvation appeared – and in the most unlikely of guises. No one could have guessed on first meeting that the scruffy-looking gentleman who arrived with Colonel Lethbridge-Stewart and called himself simply 'the Doctor' was the genius who would ultimately save us all. No one except Professor Travers, who knew the Doctor, together with his two unlikely companions, a young Scotsman and a pretty girl, from a previous encounter. But more of that later.

Suffice to say that I for one was sceptical of the Doctor's abilities. Looking back now, I suspect my eagerness for information, my enthusiasm for a 'story' may have been something of a distraction to the military/scientific team on whose metaphorical shoulders the fate of the entire world now rested.

Colonel L-S went over the time line of events, using slides to accompany.

5 February – Fog first appears. Limited and sparse, but spreads rapidly.

6 February – First disappearances. Fog itself seems non-toxic. But people who go into it don't always come out again.

7 February – Fungus sighted in Underground Tube system abandoned. The 'web' seems more dangerous than the fog, smothering victims.

9 February – First Yeti sighted. Discussion about possible change in appearance – 'a Mark II' perhaps, according to Dr. (Like Travers he has prior experience.)

Currently – Fog now covers area enclosed by Circle Line. Impervious to chemicals, flame-throwers, explosives. Discussion of possible courses of action and recommendations.

★ ★ ★

Ma...

A SINGULAR MYSTERY

THE RESIDENTS OF CAVERSHAM STREET WERE THIS WEEK PRESENTED WITH THE MOST BIZARRE MYSTERY. LAST WINTER, THE CHILDREN LIVING IN THE AREA, IN CONCERT WITH URCHINS AND WORKHOUSE CHILDREN, CONSTRUCTED SEVERAL SNOWMEN. RESIDENTS AT THE TIME REMARKED ON THE SKILLED WORKMANSHIP OF THE YOUNGSTERS AND CONGRATULATED THEM ON THEIR CONSTRUCTION.

Now it seems the snowmen have returned. Overnight, several snowmen appeared on the pavements of Caversham Street. While this is not in itself remarkable, residents are certain that the snowmen are extremely similar in their construction, and even their position, to the snowmen constructed last winter.

Since this singular mystery was first made public, residents of other areas have come forward with similar claims pertaining to the snowmen in their own streets and squares. So far there seems to be no explanation for this bizarre and unsettling turn of events.

One version of 'The Whisper Men', a traditional rhyme thought to originate in Victorian London.

Do you hear the Whisper Men?
The Whisper Men are near.
If you hear the Whisper Men,
then turn away your ear.
Do not hear the Whisper Men,
whatever else you do.
For once you've heard the
Whisper Men, they'll stop and
look at you.

Whisper Men were apparently sighted in London in 1893. Reports that they were seen in the company of Dr Walter Simeon have cast these sightings into doubt, as Simeon died in December 1892.

My Dearest Clara.

The Doctor entrusted me with your contact details in the event of an emergency, and I fear one has now arisen.

Assuming this letter will have reached you as planned, on April the tenth, 2013, please find and light the enclosed candle. It will release a soporific which will induce a trance state, enabling direct communication across the years.

However, as I realise you have no reason to trust this letter, I have taken the liberty of embedding the same soporific into the fabric of the paper you are now holding.

Speak soon.

Vastra

While information about the Whisper Men is vague and often contradictory, quite a lot is known about the robot Yeti. From UNIT files, and the writings of Travers, tougher with the suppressed articles produced by Harold Chorley, we can piece together a good deal.

The Yeti, or Abominable Snowmen, are the robot servants of a disembodied Intelligence (which may itself be linked to – or even somehow derived from – Dr Walter Simeon). Each Yeti was controlled by means of a metal sphere, which resided in a cavity in the creature's chest. This control sphere, the UNIT files speculate, contained a fragment of the Intelligence itself – enough to animate and control its robot servant.

Information in the files also implies that Travers was able to reverse-engineer a sphere and so gain control of a Yeti. There is also speculation that reactivation of a control sphere might have been what triggered the return of the Intelligence – providing it with a link to reality and a means of reanimating the Yeti in Silverstein's collection.

In his account of his time in Tibet, Travers describes how the possessed Master of Det-Sen Monastery – Padmasambhava – used figurines of the Yeti to control them. Moving these figurines on a representation of the actual landscape outside the monastery enabled the Intelligence, through Padmasambhava, to direct them to where they were needed. Similarly, the UNIT account of the London Menace incident describes how the same figurines, brought back from Tibet by Travers, were used to guide Yeti to their targets.

It seems clear from all accounts that it is unlikely anyone can survive a direct attack by a Yeti. But removing the control sphere from an inactive Yeti will render it dormant. If you do encounter a Yeti, since they are controlled and directed by the intelligence, the best option may well be simply to keep out of the creature's way. Unless you are its specific target, it will probably ignore you and continue with its designated mission. But this is by no means guaranteed…

There has been no report of any Great Intelligence activity in recent years, although it has been tentatively linked with several incidents. While it was defeated and apparently banished from this dimension at the conclusion of the London Menace, it is always possible that the Intelligence might re-establish a link with reality and return.

BEHIND THE SCENES

Although it was set in the Himalayan foothills of Tibet, location filming for *The Abominable Snowmen* actually took place in the cold and bleak, but less snowy, countryside of the Nant Ffrancon Pass in Snowdonia, North Wales. The weather wasn't kind on the week-long shoot, with the Yeti occasionally being blown over.

The Yeti costumes, designed by Martin Baugh, were made of fake fur and constructed over a bamboo framework padded out with foam rubber to give the fur structure and shape. The reasoning behind the design was that living in a cold climate, the Yeti would be covered either with fur or feathers. Fur was easier to work with. The fur was given some texture using black car paint, with the darkened 'face' a cross between a bear and an owl. The feet and claws were moulded from rubber.

For the 'real' Yeti that Travers glimpses at the very end of the story, the padding was removed from one of the Yeti costumes, making it seem less bulky and monstrous.

A flap of fur on the creature's chest concealed the cavity where a lightweight control sphere prop could be inserted. Most of the spheres were simply hollow silver balls. But one was constructed on a children's toy vehicle, so that it could 'roll' along the ground.

The central role of British explorer Travers was

THE SECOND DOCTOR

THE ABOMINABLE SNOWMEN
Written by Mervyn Haisman and Henry Lincoln
Directed by Gerald Blake
First Broadcast: 30 September–4 November 1967

THE WEB OF FEAR
Written by Mervyn Haisman and Henry Lincoln
Directed by Douglas Camfield
First Broadcast: 3 February–9 March 1968

THE ELEVENTH DOCTOR

THE SNOWMEN
Written by Steven Moffat
Directed by Saul Metzstein
First Broadcast: 25 December 2012

THE NAME OF THE DOCTOR
Written by Steven Moffat
Directed by Saul Metzstein
First Broadcast: 18 May 2013

```
                    THE ABOMINABLE SNOWMEN
          (Tracking shot following large footprints
                   Tracks lead into trees.
   Follow them until we pick up and see the foot of a yeti
   hiding behind a tree, slowly the camera starts to pan up
   the side of the body. A large hand comes into view. The
   camera continues to pan slowly up. Suddenly we hear the
           sound of the doctor approaching. (Whistling?)
```

played by Jack Watling – the father of Deborah Watling, who played the Doctor's companion Victoria at the time.

The *Doctor Who* production team was so impressed with the scripts for *The Abominable Snowmen* and the concept of the Yeti that a sequel was commissioned from writers Mervyn Haisman and Henry Lincoln even before the story had been transmitted. But for *The Web of Fear*, broadcast just three months later, the Yeti would undergo a make-over…

BEHIND THE SCENES

Location filming for *The Web of Fear* took place rather closer to home – in London, with the memorable set-piece battle in Episode 4 between soldiers and Yeti taking place in and around Covent Garden.

But one place the production team did not film was in the London Underground, as this would have been too expensive. Almost all the Underground scenes were shot in the BBC's Lime Grove Studio D. The sets, designed by David Myerscough-Jones, were constructed such that elements could be repositioned to form different tunnels and sections of station. The end result was so realistic that London Transport believed the BBC had actually filmed in on their premises without permission, and complained.

Director Douglas Camfield felt that the Yeti in *The Abominable Snowmen* were just a bit too cuddly. So for their second story, the Yeti were redesigned, although an 'original' Yeti does appear in Silverstein's museum before changing into the new version as it comes to life. Martin Baugh redesigned the costumes for *The Web of Fear* making them slimmer and adding glowing eyes to the heads. The eyes, provided by the BBC's Visual Effects Department, meant the Yeti could be seen in the darkness of the Underground tunnels. The flare from the lights also masked the details of their shape, adding to the sinister effect.

The result was one of the most memorable and frightening *Doctor Who* stories of the 1960s.

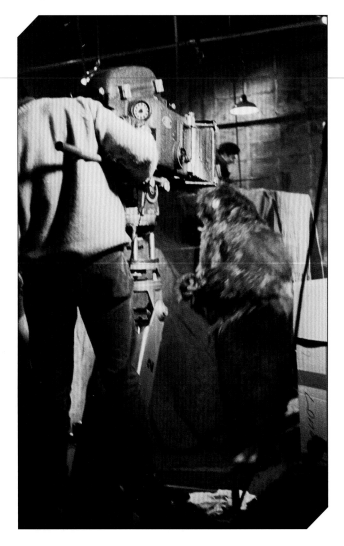

BEHIND THE SCENES

Although a third Yeti story was planned for the Second Doctor's final season, it was never made. A lone Yeti did appear briefly in the 20th-anniversary story *The Five Doctors*, but there was longer than the forty-year gap between the narrative events of *The Abominable Snowmen* and *The Web of Fear* before the Great Intelligence properly reappeared to do battle with the Doctor.

But this was a very different take on the Intelligence – with *The Snowmen* providing an insight into the origins of the Great Intelligence in the tragic figure of Dr Walter Simeon. Simeon's abominable snowmen were very different from the Yeti, being actual snowmen animated by the Intelligence in Victorian London.

While static prop snowmen were used, the fanged faces and the moving/appearing snowmen were created digitally as Computer-Generated Images (CGI). The ice governess that threatens the Doctor, Clara and the children in her care was also CGI.

When the Great Intelligence returned again in the form of Dr Simeon (following a brief appearance in *The Bells of Saint John*) in *The Name of the Doctor*, he was assisted by the sinister Whisper Men. It was a combination of costume, prosthetic make-up, and CGI that brought these menacing figures to life.

With Clara foiling the Great Intelligence's plan to unravel the Doctor's past lives, it seems that the Intelligence has finally been defeated. But it has been defeated before, and that hasn't stopped it returning… ●

THE SNOWMEN
Crump! Crump! Crump! Snowmen, all shapes and sizes, are extruding out of the snow all around them.
The terrible mouths now yawning open - those terrible icy fangs.
The workmen looking around, starting to panic - what the hell is this??
One of them tries to run - a snowman shoots up in front of him, a dazzle of ice fangs.
Now screams and panic everywhere - cutting fast round fanged mouths and flailing workmen.

Ice Warriors

THE FIRST REPORTS of a mammoth found frozen in ice date back to 1692, with carcasses excavated from the ice in Europe as early as 1728. While some bodies were remarkably well preserved by the freezing – in some cases they still had blood in their veins – there was no question of a mammoth actually surviving the process and being revived when thawed. But by the early twenty-first century, scientists had successfully thawed out frozen viruses and found they were still, in effect, alive. What if a creature existed that could survive being frozen in the polar ice for thousands of years – a creature, perhaps, from another world?

In 1983, the Soviet nuclear submarine *Firebird* suffered a near-catastrophic breakdown in Arctic waters. Details of the incident have never been made public, and at the time the Russians played down any possible risk. There were fatalities, though the actual number is unknown, but the surviving crew managed to repair the submarine's systems and bring it safely to the surface.

What is known is that *Firebird*, in addition to its Cold War duties, was on a scientific mission to investigate a particular area of polar ice. On board was the eminent Soviet scientist Professor Grisenko. After his death in 2011, Grisenko's personal papers were donated to the Library of St John the Beheaded. They included notes he had taken on the *Firebird*'s fateful voyage back in 1983. His journal details how he and his team excavated what they at first believed might be a baby mammoth. But on closer inspection it turned out to be a two-legged creature, standing over seven feet (two metres) high and of reptilian appearance.

And most strikingly, according to Grisenko's notes, the 'Ice Warrior' – as he refers to it – was still alive. He describes, briefly how his assistant Piotr disobeyed his orders and thawed the ice surrounding the creature with a blowtorch, releasing it. What happened after that is a mystery, but it seems likely that the thawed Ice Warrior was somehow responsible for the deaths and the damage to the submarine. Grisenko's sketches of the creature give a fascinating glimpse of an alien life form.

How can we be sure that this Ice Warrior was actually an alien and not some extinct form of Earth life? UNIT certainly took an interest in the *Firebird* incident at the time, and NATO reported tracking an unidentified aircraft in the area which appeared from nowhere and then simply vanished from RADAR screens.

Amongst the leaked UNIT files, there are some documents which have been largely denounced as fakes. These include a bundle of notes and documents labelled *TARDIS Information Files*. The reason these are usually discounted is because they detail events not of the present or the past, but of the future. These files also mention 'Ice Warriors'. They discuss the discovery of Ice Warriors frozen in the advancing glaciers of a future ice age that will engulf Europe. They also detail an attempted invasion of Earth by the Ice Warriors who tried to change the atmospheric composition of our planet to match that of their own original home – Mars.

The Federation assessment committee meets High priest Hepesh

In UNIT's Ice Warrior files, there are also extensive details about a planet called Peladon, and its relationship with something known as the Galactic Federation.

This Federation seems rather like a combination of the European Union and NATO, an alliance of mutually trading worlds sharing economic and military resources. The Ice Warriors were among the races signed up to this Federation, along with Earth, Arcturus, Alpha Centauri, and Vega Nexos amongst others.

Peladon is described as a feudal planet, ruled by a monarch and an aristocracy. Religion is important, with the High Priest of local deity Aggedor — a creature that is also the 'royal beast' — wielding significant political power. In fact, the role of High Priest and Chancellor was combined at some point between the two main incidents discussed in the files.

The first of these incidents is Peladon's joining of the Galactic Federation. A committee of delegates from other member races evaluated Peladon as a prospective member. One of these was an Ice Lord — a senior Ice Warrior — named Izlyr, who attended along with his adjutant Ssorg. Despite attempts by local objectors to sabotage the process, Peladon was indeed admitted to the Federation.

A rare image of the interior of the Temple of Aggedor

King Peladon of Peladon greets an Earth Princess

Queen Thalira of Peladon

Fifty years later, with King Peladon now succeeded by his daughter Queen Thalira, the people of Peladon must have wondered if they made the right choice. The Federation was at war with Galaxy Five, and Peladon became one of the main suppliers of trisilicate – a rare ore vital to the war effort.

With unrest amongst the trisilicate miners and rumours of the Federation incurring the wrath of Aggedor, the Federation Ambassador on Peladon called in security forces. But the Ice Warriors who arrived to restore order, led by Lord Azaxyr, seem to have been working clandestinely for Galaxy Five. It is interesting, reading between the lines, to see how some factions of this race of warriors rebel against the more peaceful and stable regime of the Federation. Azaxyr represents those Ice Warriors who want to see a return to the days of 'death or glory' – the days when they tried to invade our own planet.

Trisilicate miners on Peladon

We have endeavoured to keep the specimen frozen, disconnecting all heating to the rather cramped area that I am allowed as a laboratory. I really do not want to thaw the specimen out until we are back in a controlled environment, preferably at the Institute in Moscow. Piotr, I know, is more impatient to see what treasure the ice hides. The impetuousness of youth!

As the ice thaws a little - it is unavoidable - so the surface has cleared slightly and we can get glimpses of what lies hidden inside. It is tall, over 2m high, standing on two legs. Green, textured, reptilian almost - is it wearing armour? An ancient warrior perhaps? But is that possible from so long ago? The helmet, if it is a helmet, looks medieval (not unlike the find at Sutton Hoo in Britain).

The fool!

Piotr couldn't wait after all. He took a blowtorch to the ice, and thawed out the creature. Now Piotr is dead and the creature has revived.

I was right - he is a warrior, and he certainly isn't human. The strange Doctor and his young friend seem to know a lot about it. All I know is - it kills.

Now it is loose on the boat, and hungry like the wolf.

Sketches and a watercolour included with Professor Grisenko's notes and papers, presumably made by Grisenko himself. Note the lizard-like texture of the warrior's armour and the slightly 'medieval' style helmet.

Let me welcome you all to Britannicus Base, the most recent Ioniser Station to be opened under the World Program.

Thanks to the efficiency of our modern civilisation under the direction of the great World Computer, we conquered the problem of world famine a century ago by artificial foods. On the land that was once used to grow the food we needed, we built up-to-date living units to house the ever-increasing population. But Nature, it seems, will not be tamed. Not completely.

As we reduced the number of growing plants on the planet to an absolute minimum, so we reduced the levels of carbon dioxide in the atmosphere. Where once we feared that too much carbon dioxide would catastrophically raise the temperature of our world, now we have suffered the reverse effect. With no heat retention, one year, there was no spring. And so we find ourselves in the grip of a new Ice Age.

As the ice caps continue to advance, the Academy of Scientists has determined that ionisation is our only hope of turning back the glaciers. Ionisation, as many of you know, is a method of intensifying the sun's heat onto the Earth, but into particular areas. It is by no means a perfect solution. Precise control is not easy, and as ionisation can produce temperatures intense enough to melt rock we need to maintain a perfect balance of ionisation progress across the world so as to prevent widespread flooding.

But I am confident that with your help, under the scientific guidance of my colleague Scientist Penley and supervised by the ever-efficient Miss Garrett, we will soon be making a decisive contribution to the World Program and the Ice Age will simply melt away.

Fragment of a radio transmission, believed to be between Varga and his second in command Zondal, prior to the sonic cannon attack on Britannicus Base

VARGA: I am now outside their base. Stand by, gun control. Take target readings.

ZONDAL: Ready. Vertical bearing minus point nine. Lateral bearing three five... Prepare to charge to frequency three five.

>> PAUSE <<

ZONDAL: The gun is fully ready, Commander.

VARGA: Good, Zondal. We will now contact the scientists. On my command you will fire once. Do you understand?

ZONDAL: Yes, Commander.

T-MAT – THE FUTURE OF TRANSPORT

Travel-Mat is the ultimate form of travel. Control centre of the present system is the Moon, serving receptions at all major cities on the Earth. Travel-Mat provides an instantaneous means of public travel, and transports raw materials and vital food supplies to all parts of the world.

T-Mat supersedes all conventional forms of transport. Using the principle of dematerialisation at the point of departure, and rematerialisation at the point of arrival in special cubicles, departure and arrival are almost instantaneous.

Although the system is still in its early stages, it is completely automated and foolproof against power failure.

T-MAT DELIVERS!

Marketing material produced to support the announcement of the Worldwide T-Mat Network

Photographs taken from security feeds in the T-Mat base on the Moon, and in and around T-Mat headquarters, London

TO: SIR JAMES GREGSON UN PLENIPOTENTIARY & MINISTER WITH SPECIAL RESPONSIBILITY FOR T-MAT

Dear Sir James

Thank you for your confidential e-memo of yesterday. I was not aware of Professor Eldred's activities, nor of the rocket you say he is constructing. As you know Daniel Eldred and I used to be friends before I took up my current position with T-Mat Control, and I speak from experience when I tell you that ordering him to cease this activity will only stiffen his resolve.

My considered opinion – and my firm recommendation – is that we simply let him get on with it. Despite his undoubted brilliance and his expertise in old rocket technology, the chances are that the vehicle will never be completed or viable. Even if it is, Eldred's project hardly poses a threat to T-Mat.

More likely, the construction forms another of Professor Eldred's exhibits for his 'space museum'. As such, I think we can safely ignore it.

Assuring you of my best wishes

Commander Julian Radnor

Mighty is Aggedor, fiercest of all the beasts of Peladon. Young men would hunt him across the slopes of Mount Megeshra to prove their courage. His fur trims the King's royal garment. His head is the royal emblem of Peladon.

But with the passing of the seasons and the moving of the stars, the royal beasts became scarce. Their numbers dwindled and fell. And there came at last a time when the roar of Aggedor was heard no more in the land of Peladon, and much was the sadness and lamentation at the passing of such a noble beast.

But it is written in the testaments and chronicles of old that there will come a day when the spirit of Aggedor will rise again. The spirit of Aggedor will rise to warn and defend his royal master, King Peladon.

For at that day, a stranger will appear in the land, bringing great peril to Peladon.

> An account of the Legend of Aggedor. The traditionalists who attempted to keep Peladon out of the Federation claimed that the wrath of the spirit of Aggedor at the acceptance of 'outsiders' was to blame for the death of Chancellor Torbis.

For Attention Of: Secretary General, the Galactic Federation
From: Madame Chairman Amazonia, Peladon Assessment Committee

Secretary General – While I wholeheartedly endorse the findings and recommendations of my fellow Delegates on the Committee and look forward to welcoming Peladon into the Federation, I must protest in the strongest terms to the manner in which these findings and recommendations came about.

As you are aware, my arrival on Peladon was delayed by a necessary diversion to Oralandis Minima to act as the Federation's appointed Arbiter in the dispute there between the Varlakkers and the Non-Aligned Krinchipians. On arrival on Peladon, I discovered that not only had my communications not been received (owing to sabotage of the other Delegates' comms systems), but my position as Chair of the Committee had apparently been usurped by an imposter.

Although this imposter and his accomplice had already fled the scene by the time of my arrival, I am astonished to learn that Peladon – and indeed the Federation itself – has made no effort to ascertain the true identity of this self-styled 'Doctor' or the young woman calling herself Princess Josephine. It seems likely that their nefarious plan was for the 'princess' to become Queen of Peladon. I'm sure I do not need to stress how embarrassing, and politically dangerous, it would have been for us had they succeeded.

I urge you to launch an immediate investigation into the incident, and alert Federation Security forces in the area so that the Doctor and 'Princess Josephine' may be speedily detained.

Regards

Amazonia

> A memo sent by the Chair of the Peladon Assessment Committee immediately after the ceremonial coronation of King Peladon of Peladon and the formal signing of the Peladonian Treaty of Accession

The evidence, therefore, would suggest that Commander Azaxyr together with his Sixth Legion has defected and formed an alliance with the forces of Galaxy Five. There has been no formal statement from the Master of the Fifth Galaxy, and Azaxyr's current whereabouts are unknown.

Intelligence reports last place Azaxyr's forces in the vicinity of Principia, on the outer limits of the Federation but well within striking distance of our strategic facilities on Mendalya, Bascam, and Peladon. Security troops have been despatched to the area, and encoded lightstream messages sent to our Ambassadors on each of these planets alerting them to the possible danger.

Only the Ambassador on Peladon has not acknowledged receipt and forces are being diverted to that planet as a priority. You will be aware of the current delicate political conditions on Peladon, and of the vital importance of our trisilicate mining operations on that world. Axazyr's intervention could destabilise the situation, and there is a possibility that he might acquire the refined trisilicate for use by Galaxy Five in their own war effort. As you are aware, their technology is as dependent as ours on trisilicate, and the latest assessment of the Committee for Strategic Affairs is that whoever controls the most stable and plentiful supply of trisilicate will almost certainly win the war.

Images released by the Galactic Federation once the crisis on Peladon was over

Extract from a Federation Intelligence report into the activities of Lord Azaxyr

From Professor Grisenko's papers and the UNIT files, we can piece together significant information about the Ice Warriors.

Originally from the planet Mars, the nickname 'Ice Warriors' seems to have been adopted by the race, and refers to their affinity with cold climates. The Ice Warriors can evidently survive being frozen, but conversely seem to be susceptible to heat.

The Ice Warrior and Ice Lord are in essence the same creature, wearing different armour. The Ice Warrior that we see is actually an armoured, cybernetic suit. Incredibly strong, resilient and robust, it is also equipped with built-in weaponry in the form of a powerful sonic disruptor attached to the right forearm.

Much of the Ice Warriors' technology is sonic-based. There is mention of their use of a sonic cannon to attack Britannicus Base during the next ice age, for example.

The distinctive rasping, hissing voice may not be 'natural' but is possibly due to the fact that the creatures prefer a different atmosphere to our own. Hence their attempt to use the T-Mat transportation system to distribute Martian seed pods around the world. These pods burst, emitting spores of a fungus-like 'blight' which affects crops and spreads rapidly across the landscape, changing the composition of the atmosphere to something closer to the Ice Warriors' needs.

It is fortunate that this race – with its militaristic aristocracy and deep-rooted sense of honour coupled with an enthusiasm and talent for warfare – will one day seek more peaceful means of influencing other races through the economic and political instruments of the Galactic Federation.

BEHIND THE SCENES

The Ice Warriors that appeared on screen were very different creatures from those envisioned by writer Brian Hayles (pictured right) and described in his script for *The Ice Warriors*. Hayles's concept was closer to human soldiers in medieval-style space armour. It was costume designer Martin Baugh who came up with the idea that the Martians should be reptilian bipeds fused into armoured shells. His initial idea was for a kind of upright crocodile with a Viking-style helmet.

Once the designs were complete, the Ice Warrior suits were made largely of fibreglass with rubber joints. Despite the joints, they were very inflexible, and had to be bolted together with the actor inside. The mouth was made up with latex to appear reptilian. A plan to make the eyes glow behind the tinted glass covers was dropped, partly because the lights might not show up under the bright studio lighting, and also because the suits were quite hot enough without putting light bulbs inside them.

The first time the costumes were used was for filmed sequences at the BBC's Ealing Studios. As a result of this, changes were made, in particular to the Warriors' heads. The original versions were rather top heavy, and allowed a very limited range of movement. To alleviate these problems, Baugh

added a flexible neck piece made from rubber and latex. The helmets were redesigned and extended downwards to cover the chin. Comparing the shots of Turoc chasing Victoria through the glacier or the sequence of Varga waking at the end of Episode 1 with the material shot later in the TV studio, the changes are apparent.

Director Derek Martinus cast Bernard Bresslaw as the leading Ice Warrior, Varga. Bresslaw was a very tall actor, best known for his frequent appearances in the *Carry On* series of films. It was Bresslaw who created the hissing, sibilant voice of the Warriors. Thinking of how reptilian the costume looked, he experimented with a hissing, lizard-like voice. It was not treated, but because of the problems of acting and speaking in the full make-up and costume, the Ice Warriors' voices were pre-recorded then played back into the studio. The actors playing the Warriors then synced their lips to their recorded voices.

The effect of the Ice Warriors' sonic disruptors was a distortion of the target as it is hit by sound waves. This was achieved by pointing the camera at a sheet of flexible mirrored plastic showing a reflection of the target. When the plastic was pushed gently from behind, so the image was distorted and wobbled.

BEHIND THE SCENES

The same costumes were used again when the Ice Warriors returned in *The Seeds of Death*. But this time their leader Slaar was not just another 'warrior' like Varga but an 'Ice Lord'. Slaar was sleeker and shorter, with a chestplate in place of full armour and a different, smoother helmet that left his mouth and chin exposed. These were heavily made up, and the teeth blacked.

Actor Alan Bennion played Slaar and both subsequent Ice Lords, Izlyr and Azaxyr. With his eyes hidden behind plastic lenses, Bennion had such limited vision that if he was the first to speak in a scene, he had to be given his cue with a tap on the back of the legs by an assistant keeping out of sight of the cameras. Bennion spoke his lines in 'real time', but again, the other Warriors' lines were pre-recorded and played back into the studio.

Bennion's costume was refined for the Third Doctor story *The Curse of Peladon*, and a cloak was added to denote the aristocratic status of Lord Izlyr. The only Ice Warrior to appear in the story was Izlyr's deputy Ssorg – played by Sonny Caldinez, who had appeared as a Warrior in both previous stories. This time, he wore the costume originally created for Bernard Bresslaw to wear as Varga.

Alongside the Ice Warriors, now allies of the Doctor, were several other alien creatures. Most memorable was the nervous one-eyed, six-armed Alpha Centauri. Another delegate, Arcturus, was a grotesque disembodied head in a transparent plastic tank on top of a mobile life-support system. Also appearing was the royal beast of Peladon, Aggedor – a large, furry animal with claws and horn.

Both Bennion and Caldinez returned for the Third Doctor's second visit to Peladon – *The Monster of Peladon* – this time as the villainous Lord Azaxyr and his second in command Sskel. Again, Bennion spoke his lines during recording while the Ice Warriors' lines were pre-recorded. But this time they were not even spoken by the actors playing

the Warriors, but by *Doctor Who* producer (and former actor) Barry Letts.

Although this was the last time the Ice Warriors appeared in the original series of *Doctor Who*, one Ice Warrior costume did feature briefly in *Genesis of the Daleks* the following year. Glimpsed through a grating, and with material added underneath it, the 'shell' was moved horizontally past the camera to represent one of the failed genetic experiments of Dalek-creator Davros.

THE SECOND DOCTOR

THE ICE WARRIORS
Written by Brian Hayles
Directed by Derek Martinus
First Broadcast: 11 November-16 December 1967

THE SEEDS OF DEATH
Written by Brian Hayles
Directed by Michael Ferguson
First Broadcast: 25 January-1 March 1969

THE THIRD DOCTOR

THE CURSE OF PELADON
Written by Brian Hayles
Directed by Lennie Mayne
First Broadcast: 29 January-19 February 1972

THE MONSTER OF PELADON
Written by Brian Hayles
Directed by Lennie Mayne
First Broadcast: 23 March-27 April 1974

THE ELEVENTH DOCTOR

COLD WAR
Written by Mark Gatiss
Directed by Douglas Mackinnon
First Broadcast: 13 April 2013

MONSTER OF PELADON

THE MASSIVE DOOR GRINDS OPEN SLOWLY. ORTRON AND THE GUARDS, COMING UPON GEBEK AND DOCTOR WHO, STOP IN THEIR TRACKS, STARING AT THE OPEN DOORWAY. STANDING IN IT IS THE GRIM FIGURE OF AN ICE WARRIOR. IT RAISES ITS MIGHTY FIST, ON WHICH IS SET ITS SONIC EXTERMINATOR.

BEHIND THE SCENES

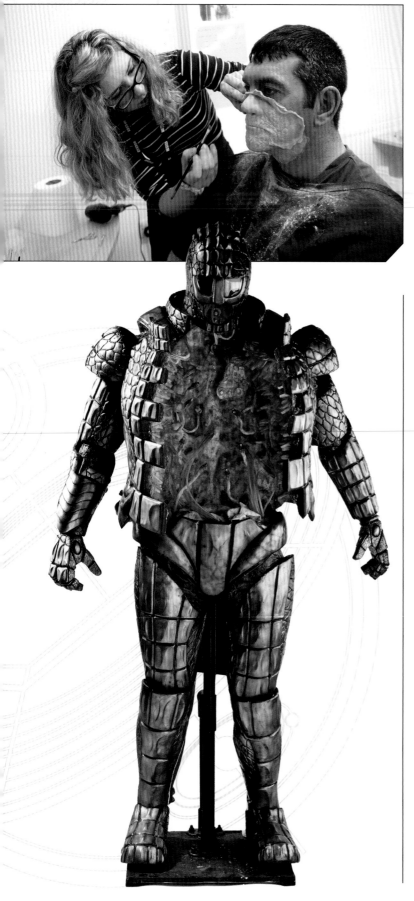

The single Ice Warrior that appeared in *Cold War* – Grand Marshal Skaldak – was very similar to the Ice Warriors that had appeared almost forty years previously. In a similar way, the story itself echoed the discovery of Varga, the first Ice Warrior, discovered frozen in ice and inadvertently brought back to life when thawed out. There were changes made to the new costume, designed and made by Millennium FX, but the overall design remained very true to the original.

The body shell was a little more streamlined, projecting less at the front, and made to look more like armour plating. The hair that emerged at the joints and neck was gone. The clumsy clamps that served as hands were replaced with a more dextrous arrangement of two stubby fingers and a thumb. The main part of the suit was made from flexible urethane rubber, which was considerably lighter and more comfortable than the original fibreglass.

Again, only the area round the mouth and chin required make-up (although the original Ice Warriors also had their eyes blackened), and Skaldak was given a scar, presumably picked up in battle.

The creature that escapes from inside the suit – the actual Martian itself – was realised partly as

```
                         COLD WAR
                       Drip-drip-drip.

                          Then —

Boom! The ice shatters and a huge, clamp-like claw shoots out, grabbing Piotr by the
                          throat!

He gasps, choking as the claw squeezes the life from him. Desperately he flails at the
              massive scaly arm but it's no use. He drops down, dead.

                  Great lumps of ice smash to the floor.

In the glare of the faulty, sputtering lights, a colossal figure is revealed. Seven
     feet tall. Green. Scaly. With a huge armoured head and clamp-like hands —
```

a model puppet in some scenes. But when seen properly as the full creature, the puppet version was not thought to be convincing or frightening enough and so CGI was used to create the creature digitally.

The physical Ice Warrior was played by actor Spencer Wilding. Again, as in *The Monster of Peladon*, the voice came from another actor – voice artist Nicholas Briggs, who also provides the voices of Daleks, Cybermen, Judoon and other *Doctor Who* monsters.

With the full effect coming from a combination of costume, make-up, performance and digital effects, the Ice Warriors are just as chilling now as they were when they first appeared out of the glaciers almost 50 years ago... •

The Judoon

FOR A FEW HOURS, in the immediate vicinity of the Royal Hope Hospital in London, the rain fell *upwards*. Reported at the time as a freak meteorological incident caused by an unprecedented collision of the Jet Stream and the Gulf Stream coupled with particularly high atmospheric pressure, this remarkable phenomenon made the front pages of almost every national newspaper. Coupled with claims that the weather also affected people's perception and caused minor hallucinations, the story only missed out on a full set as one 'red-top' tabloid chose instead to report that prominent politician Harold Saxon had allegedly declined to appear on TV reality show *Big Celebrity Dance Mania*.

The alleged hallucinations themselves received scant attention, lost in the British fascination with the weather in general and wet weather in particular. But it is surely not just a coincidence that so many people in the area saw the same thing. Or rather, failed to see it. According to their accounts, for several hours the entire Royal Hope Hospital *simply* vanished.

The stories told by the people inside the hospital at the time of this freak weather are also bizarre. But, again, they are consistent with each other. The witnesses claim the whole facility was transported to the surface of the Moon. They tell of sinister bikers, complete with visored helmets. They talk about an elderly female patient who apparently fed on the blood of other patients and even a senior consultant. And they tell of the Judoon – intergalactic policemen that look like upright rhinoceroses dressed in dark leather uniforms who dispensed lethal 'justice' at the slightest provocation.

A collective mass hallucination brought on by freak weather conditions? Or something altogether more disturbing?

Galactic criminal, or just making a delivery?

An artist's impression of what a hospital might look like if transported to the Moon

St. Thomas' Hospital

ACCIDENT & EMERGENCY

Galactic police, or mass hallucination?

Note the distinctive helmet shape.

Could this be what's beneath that helmet?

With all security camera footage from the Royal Hope Hospital strangely missing – apparently due to a malfunction in the CCTV systems caused by the weather conditions – there is no absolute proof of what may have happened during those few hours the hospital was apparently 'missing'. But there is enough agreement between the people involved, as well as other evidence including photos purportedly taken on mobile phones, to support the theory that the hospital was indeed removed from the Earth's surface and visited by aliens.

What can we glean about these supposed aliens from the witnesses' reports? Everyone agrees that they wore dark, leather uniforms with strangely shaped helmets that jutted out at the front. A large number of people also claim to have seen at least one of the aliens without its helmet – and say that it had the head of an upright rhinoceros, complete with horns. The name 'Judoon' was overheard by many.

One theory, again supported by the witness statements, is that these Judoon have no jurisdiction on Earth and so engineered the removal the of the hospital to the 'neutral territory' of the Moon in order to dispense their galactic justice. There is broad agreement that the Judoon were searching for a criminal concealed among the patients and hospital staff, and subjected everyone to an analysis to see if they were indeed human.

Medical student Oliver Morgenstern was working at the hospital that day, and his account is one of the more detailed. Morgenstern says that the Judoon were looking for a blood-dependent 'Plasmavore', which had fled from justice after killing 'the Child Princess of Padrivole Regency Nine'. That's as may be, but it does seem that an elderly Miss Finnegan was being treated for a salt deficiency. The NHS number under which she was treated does indeed belong to one 'Florence Finnegan' – who died over 30 years earlier in 1974. So just who was the woman at the hospital that day? She has never been seen since.

Spacecraft on the Moon — fact or fiction?

Was this woman the mysterious Miss Finnegan?

Huge leather-clad creatures stomped across the foyer, marching in step like some bizarre battalion of soldiers just back from yomping across the Welsh mountains rather than the lunar craters.

The leader stopped right in front of me, my head level with his chest. I looked up at the opaque mask of his helmet, a strange jutting shape that protruded out from the body. I could feel the fear growing inside me like an inflating balloon, ready to burst at the prick of a pin. The other staff gathered behind me, clearly seeing me as their reluctant spokesman.

If only I *could* speak. My voice seemed cleft to the back of my throat. A situation that only worsened as the behemoth reached up and unfastened its helmet. With a hiss that might have been that balloon in my chest rupturing, the clasps released. Despite what I had already seen, nothing could have prepared me for what I saw beneath the helmet. The wrinkled skin, like old cork. The eyes set deep in the face – if you could even call it such. Ears so small they could have been the afterthought of some cosmic designer of monstrosity.

And horns!

Behind me, I heard the horrified gasps of my colleagues. The embarrassing sound of Hazel Magee fainting dead away at the horrendous and impossible sight. My first thought, though, despite all this, was

An extract from Oliver Morgenstern's novel based on the events at the Royal Hope Hospital: *Accidental Moon*. It was not kindly reviewed.

THE WORLDS BE... ...AL NEWSPAPER

Est - 1965

Monday 5th June

The tabloid newspapers ignored the bigger story...

WHAT GOES UP?!

Londoners were treated to a unique experience yesterday when the rain went backwards.

A freak storm over the Royal Hope Hospital was, the Met Office say, the result of a one-in-a-billion combination of bizarre circumstances which created an updraft powerful enough for rain to actually fall upwards into the sky.

The area was soon cordoned off by police,

in case the weather conditions worsened. Patients and staff were confined to the hospital for the duration of this weather emergency.

'Only in Britain,' said a Met Offic[e] spokeswoman, 'could something like th[is] happen...'

From the UNIT files — are the Judoon recruiting on planet Earth?

THE LAW NEEDS YOU

121

(CONT. FROM PREVIOUS)

A less serious but still troubling issue is the confusion surrounding the patient named in records as 'Florence Finnegan (Miss)'. As mentioned earlier (Section 2, para 4), the patient does not appear to have survived the incident. In fact, there is some debate among witnesses as to whether she was ever actually at the hospital at all. Her named consultant was Mr Stoker, who is obviously not available for clarification. But the Hospital Number allocated to her cross-references NHS and National Insurance numbers that refer to a Florence Finnegan listed as deceased in 1974.

This might be put down to an incorrect assumption of death. But since the deceased Miss Finnegan died at the age of 93, it is inconceivable that this could be the same woman. The question therefore remains – if she ever actually existed, who was she?

And did she play some role in the death of Mr Stoker, which remains – despite all the other bizarre allegations and reports pertaining to this incident – unexplained?

An extract from the Royal Hope's own report into the incident

These pictures were posted online after the Royal Hope incident by various staff and patients who claim to have taken them on their mobile phones.

...the telegraph communication with the Shadow Proclamation following this latest incursion led to a more considered and satisfactory response. The organisation has removed jurisdiction from the security forces for the whole of planet Earth.

What this means in effect is that the Torchwood Institute is now responsible for all on-Earth transgressions. In extremis, either Torchwood or UNIT may call on the help and resources of the Shadow Proclamation as, when and if necessary, but the Judoon will no longer have carte blanche to operate... without express permission jointly agreed by bot...

An extract from just one potentially relevant document from the leaked UNIT files. Note the mention of the Torchwood Institute.

122

TO: USUAL OUTLETS
ISSUED BY: HOME OFFICE PRESS SECRETARY
FOR IMMEDIATE RELEASE
SUBJECT: INCIDENT AT ROYAL HOPE HOSPITAL

As has been made clear in previous press releases, the Home Office and the Ministry of Health have been jointly investigating the incident last week at the Royal Hope Hospital. It would be improper to comment in detail before the police have completed their own investigations and the CPS has determined if any charges are to be brought with relation to the deaths and disappearances that have been widely reported.

However, in the interests of maintaining calm and perspective, the Secretary of State has agreed to make public the Senior Health Consultant's conclusion that the majority of reports, including the most bizarre, stem from the witnesses suffering a collective hallucination caused by oxygen starvation. This in turn was itself due to the freak weather conditions in the area which, as we know, caused rain to fall upwards into the sky and which also severely depleted the oxygen levels in the lower atmosphere across a limited area, covering the Royal Hope Hospital and its immediate surroundings.

Police Report 1398101CC/15 (CONT.)

The initial assumption was that Mr Stoker had suffered from asphyxiation, in common with many of those in the facility, and that this was the cause of death.

However, examination by the forensic pathologist revealed puncture marks that were inconsistent with this diagnosis. Subsequent internal analysis and testing of organs and fluids revealed that the victim had been largely drained of blood. What plasma remained was severely depleted of both salt and fat content. The cause of death was therefore determined to be extreme desanguination coupled with desalination.

The inevitable conclusion from this is that Bryan Stoker's death was homicide – he was murdered by person or persons as yet unknown.

The police report into Stoker's death refers to 'persons'. But was his killer even human?

Martha Jones — medical student, or something more sinister?

Smith and Jones — are their names even plausible?

WHO? WHO?
Who is John Smith?
MARTHA JONES?
WHO?!?

A close look at the records of the Royal Hope Hospital for the time of the alleged incident is instructive. As well as the deceased Florence Finnegan, two other names of interest are to be found.

One of these is John Smith – an unremarkable name, but one linked to an invalid NHS number. Follow the evidence still further, and it leads to a portfolio of medical data that has clearly been manufactured. There are blood pressure and body temperature readings that are impossible for a human being, and even chest X-rays that seem to show two hearts. If the intent is to cover up the real medical condition of the patient, then it is clearly inept.

Also present at the Royal Hope was a medical student colleague of Oliver Morgenstern named Martha Jones. She is remarkable partly for not giving any sort of witness statement to the police or the press after the event, and also for taking an unscheduled leave of absence immediately afterwards. It is perhaps not surprising that when she reappears, Martha Jones completes her medical training and then joins the shadowy UNIT organisation. Could it be that UNIT was somehow involved in the Judoon incident?

And what of the mysterious motorbike couriers, dressed entirely in black leather, who seemed to be helping the alien Plasmavore in her attempts to evade justice? How many aliens were there at the hospital that day?

The Judoon themselves disappeared from the scene shortly before the hospital was apparently returned to its terrestrial location, the staff and patients suffering from oxygen starvation. Morgenstern and others report that the Judoon arrived and left in huge cylindrical spaceships, the size of office blocks. Their armoured suits seemed to protect them from the vacuum of the lunar surface, and they obviously deployed some sort of force field to prevent the air in the hospital from dissipating immediately on arrival.

John Smith?
Who IS he?!
WHO?!?
WHO?
WHO?

FURTHER INFORMATION

Many theories have sprung up in the wake of the Royal Hope Hospital incident. Most of them have little basis in fact, or in the witness accounts of what actually happened. But, following up on the UNIT connection, we can extrapolate from information in the confidential files that have become available over the last few years. Together with the evidence from witnesses, particularly Oliver Morgenstern, it provides some detail about the Judoon.

LANGUAGE: The Judoon native language consists of a guttural series of staccato exclamations. But they have the technology to sample the speech of other races with a hand-held device and synthesise a translation.

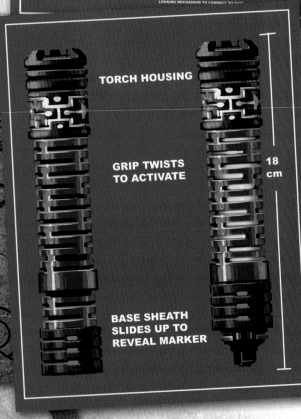

TRANSLATOR DEVICE

EXAMINATION LIGHT

WHEN THE TRIGGER IS HELD, BOTH THE MAIN'TORCH' LIGHT AND THE 'TRANSLATING SCREEN' LIGHT UP

LOCKED 21.07.06

PRAC TRIGGER ACTIVATES PROP LIGHTS

WHEN THIS SMALL BUTTON ID SLID UPWARD, THE TWO ARMS AT THE TOP OF THE PROP SWING OUT

16cm

LOCKING MECHANISM TO CONNECT TO SUIT

TORCH HOUSING

GRIP TWISTS TO ACTIVATE

18 cm

BASE SHEATH SLIDES UP TO REVEAL MARKER

TECHNOLOGY: In addition to this translator device, the Judoon have handguns. A genetic detector device was used at the Royal Hope to identify subjects as 'category human'. Once identified, the device could be used to put a mark on the back of the hand – a mark which could, presumably, not be copied or forged.

JURISDICTION: It seems that the Judoon are a mercenary police force – not unlike a contract security company, but obviously with a far wider remit. Other races in effect subcontract their security and policing duties to the Judoon. They also provide security for intergalactic non-aligned organisations such as the mysterious Shadow Proclamation (which is mentioned several times in the UNIT files).

SURVIVAL GUIDE

As the Judoon are guardians of justice, the simplest and most obvious way to avoid incurring their wrath is to stay within the law. They appear to be extremely literal – and draconian – in their administration of justice, so it is vital to adhere to the letter of the law as well as its spirit.

But our greatest defence seems to be the question of jurisdiction. From data in the UNIT files, and from the Royal Hope incident, it is clear that the Judoon are not authorised to operate on planet Earth. Remember, though, that this is a race that itself observes the letter of the law and not its spirit – that had no qualms about transporting a whole hospital to the Moon so that the protection afforded by being on Earth no longer applied.

The Judoon enforce the law as they see it, and they do so ruthlessly.

BEHIND THE SCENES

Smith and Jones was the first story of the third series of *Doctor Who* after its return to BBC One. As well as kicking off the 2007 season, the story also introduced a new companion, Martha Jones. For her introductory story, lead writer and executive producer Russell T Davies created a new race of aliens – the Judoon.

His script describes the first appearance of this galactic security force as they march out from their spaceships across the lunar surface: '... burly, tough figures in black, heads in black helmets. And armed. Like stormtroopers...' The reveal of how they look beneath their helmets was held back for a later scene.

But there is a hint as soon as the Judoon arrive, when the script describes their helmets as 'the strangest shape, like the thick head juts forward, onto the chest, then twists up at the end...' Then the Judoon Captain removes his helmet – to reveal 'The head of a rhino. A humanoid rhino. Grey leathery skin. Snout curving down then up into a horn. Helmets modelled around this...'

The Judoon were designed and created by Millennium FX and costume designer Louise Page, working from the descriptions

SIGNIFICANT SCREEN ENCOUNTERS

THE TENTH DOCTOR

SMITH AND JONES
Written by Russell T Davies
Directed by Charles Palmer
First Broadcast: 31 March 2007

THE STOLEN EARTH
Written by Russell T Davies
Directed by Graeme Harper
First Broadcast: 28 June 2008

> They stamp in, spread out, taking control of the foyer, the JUDOON CAPTAIN standing centre —
> THE JUDOON are big and brutish, stocky, thickset, in uniforms of studded black leather panels; the lower half is a leather skirt, like Roman centurions; hefty boots below. Heads are covered with shiny black helmets, in the strangest shape, like the thick head juts forward, onto the chest, then twists up at the end. It's hard to work out, until the JUDOON CAPTAIN twists a clasp at his neck — the hiss of depressurisation — and lifts off his helmet.
> The head of a RHINO. A humanoid rhino. Grey leathery skin. Snout curving down, then up into a horn; helmets modelled around this (all other Judoon keep helmets on).

in Russell's script,. Eight Judoon costumes were made but, to save on costs, just one animatronic head was produced – for the Judoon Captain, the only one to remove his distinctively shaped helmet. The head, along with the torso with its armoured plates and the helmet were designed and produced by Millennium FX, while Louise Page created the rest of the costume.

After initial design drawings were agreed, the head was sculpted in clay by Martin Rezard before being cast in a lightweight latex-like material. This was then fitted over a metal frame that contained the remote-controlled animatronics that operated the various facial movements.

BEHIND THE SCENES

Two remote controls were needed to operate the head. One control system worked the eyes, ears and nostril twitch. Others made the jaw open and shut as well as providing a 'slew' movement and a curl of the lips to form a pout or a snarl. With all the technology built into it, the final mask was quite heavy, and got very hot inside.

The distinctive voice was provided by Nicholas Briggs, better known for voicing the Daleks and the Cybermen.

The Judoon were immediately popular, and returned in minor roles in several Doctor Who stories such as *The End of Time* and *A Good Man Goes to War*. In *The Stolen Earth* it was revealed that the Judoon provide security for the powerful and influential Shadow Proclamation. Like the Sontarans and Slitheen, the Judoon even made the crossover to appear in several episodes of spin-off series *The Sarah Jane Adventures*. ●

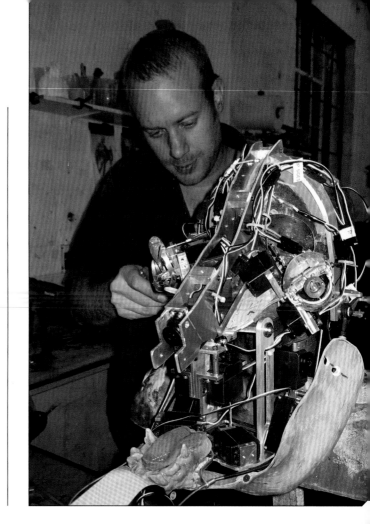

The Krillitanes

DEFFRY VALE HIGH SCHOOL hit the headlines when a substantial part of the main building was destroyed in a suspected gas explosion. The accident – which has never been satisfactorily explained – resulted in the deaths of several of the senior teachers as well as a number of the catering staff. It was a tragedy mitigated only by the fact that no children were injured.

But the truth is that Deffry Vale was attracting attention even before the accident. Several children who attended from the local orphanage had gone missing in the weeks before the explosion, although there is nothing to suggest that there is any connection. Deffry Vale had been a very ordinary school, with very ordinary exam and SATS results – until the arrival of a new head teacher.

When Hector Finch took over the school, one of his first challenges was to find temporary replacements for a number of teachers who had fallen foul of an unusual flu epidemic. Or so the story goes. He also replaced the caterers with an outside contract company. Within just a few weeks, the school's results were rocketing, and ordinary high school Deffry Vale was top o f the local authority's 'league table'.

On the face of it, this would seem to be an uplifting story about a talented head teacher who inspired his staff and pupils to attain new heights of achievement. That was certainly how the education authority portrayed it. The fact that the school's academic standing returned to its previous very ordinary status almost as soon as Finch was gone – a victim of the gas explosion – seems to bear this out. But were things really as impressive as they seemed? Consider the following:

No evidence was ever found to support the theory that the explosion was caused by a gas leak. None of the new teachers employed by Finch had worked for the local authority before – and in fact there is no record of any of them ever working as teachers in the UK prior to their arrival at Deffry Vale. The catering company that took over the school meals had no other contracts, and was set up just the week before it started work at the school. It ceased to trade the day after the explosion. NHS records show that there was no flu epidemic at the time the teachers were taken ill, and none of the teachers who left have since been located. The only exception is one female teacher whose story that she was retiring immediately as she had won the lottery turned out to be true – despite the fact her friends claimed she never played the lottery…

Perhaps most telling of all, there is no record of any Hector Finch ever registering as a teacher with the local authority, or graduating from any teacher training course at a time that coincides with Mr Finch's age and references. Not only that, but the local education authority was unable to find any paperwork relating to Finch's appointment.

So what really happened at Deffry Vale?

Is this how the teachers at Deffry Vale *really* looked?

Deffry Vale High School — exceptional academic results, or something more sinister?

Hector Finch, head teacher at Deffry Vale. Who was he really?

Soon after the explosion at Deffry Vale High School, rumours relating to events at the school began to circulate on the internet. The original source seems to have been an 'alien conspiracy' website originally set up by a man called Clive Finch (see page 148) – could this be a coincidence? Or do the bizarre claims made on the site (now, admittedly under new 'ownership') have some grain of truth in them?

According to the most persistent of the rumours, Finch and the new staff were all aliens who disguised their real appearances. Their motive in effectively taking over control of the school was to increase the mental acuity of the pupils, and then to use the children as 'organic computers' to solve otherwise insoluble problems and equations.

If this seems far-fetched, the stories go on to claim that the children's mental abilities were enhanced by means of additives introduced by the new caterers – many of whom were also aliens – into the cooking oil used in the school kitchens.

An artist's impression of what the bat-like alien Krillitanes might look like

Reports of a 'robot dog' seem far-fetched

The explosion that destroyed a large part of the school was triggered by an accident involving the cooking oil, and resulted in the deaths of the alien creatures including their leader, Hector Finch.

Is any of this story plausible? Perhaps surprisingly, the answer is 'yes'. Unusually for a theory of this kind, it does in fact explain all the strange events and occurrences at Deffry Vale. And even before Finch's arrival, there were reports in the local press of huge bat-like creatures being seen in the area. Those reports ceased immediately after the explosion. There were also reports of a 'robot dog' being sighted – an instance perhaps of credible and genuine reports leading to more fantastical claims...

Was Deffry Vale High School infiltrated by aliens who could somehow change their form from giant bats into surrogate humans? Was the mysterious Hector Finch, head teacher at the school, really an alien 'Krillitane' called Brother Lassar?

Since the explosion at the school seems to have destroyed the aliens and all evidence of their existence, the truth can never be known...

Alien caterers serving meals that can improve your exam results?

Dear Parents / Guardians

It is with great sadness that I have to tell you that Mr Reislip will not be returning to Deffry Vale for the foreseeable future. His family problems have proved more intractable than anticipated and so there has been no alternative but to appoint a replacement.

As you know, I have been acting Head Teacher for the past week, and I am pleased and proud to inform you that the Governors have now seen fit to make my appointment permanent. I should like to take this opportunity to thank them for their continuing support, and obviously I hope that the members of the governing body who were not able to attend the meeting last night will soon have recovered from the flu – the same flu that has struck down so many of the fine teaching staff at Deffry Vale.

You may rest assured that your child's education, health, and safety are my prime concerns at this difficult time of transition. I shall be working closely with our new kitchen staff, provided by Krillitane Caterers, to ensure that the school dinners remain as nutritious and appetising as ever. Our new computer suite is due to open next week. And I can assure you that I have personally selected the new members of the teaching staff who start work on Monday.

Together, I am sure we shall make Deffry Vale a school that will attract attention not just within the local area but far, far wider afield. I look forward to working with you all for many years to come.

H. Finch

Hector Finch

Head Teacher

A lucky winner, or something more sinister?

WEDNESDAY
MAINS
(All served with Chips)
Battered Fish
Chicken Strips
Vegetable Lasagne
Ham and Egg

SIDES
Onion Rings (3)
Extra Chips

DESSERT
(All served with custard)
Fruit Crumble
Spotted Dick
Fresh Fruit Salad

LOCAL WOMAN'S NUMBER'S UP

Lady Luck was smiling on one local woman this week when Lucille Beneluct won the jackpot in the National Lottery Thunder-Millions Draw. What's even more surprising is that Miss Beneluct (37) never bought a ticket!

"It's extraordinary," Miss Beneluct told our reporter. "Someone pushed the winning ticket through the letterbox, sealed in an envelope with my name on it and the business card of a local travel agent. I've no idea who it was, but I owe them – literally – a fortune!"

Miss Beneluct wasted no time putting her winnings to good use. As well as buying a new basket for Padsy, her Scottish Terrier, she has resigned from her job as a teacher at Deffry Vale High School and booked a Mediterranean cruise. "I'll miss the children, of course," she said. "But not much."

Pupils questioned about the caterers after the explosion all claimed that the chips were especially good.

GIANT BAT AT LARGE IN DEFFRY

Experts agree that this photo shows an actual bat, made to seem incredibly large because of the angle and perspective

Two more sightings of the so-called Giant Bat of Deffry were reported to police in the past week, bringing the total number of incidents to seven. Mrs Irene Smuthers reported seeing a "huge flying thing, like a dirty great bat" over her garden shed on Wednesday night. "I was putting out the bins, and there it was – large as life and twice as unnatural. Huge it was."

The second sighting was reported by Jerry Atkinson, a local pensioner. "I was just checking on my dahlias the other evening and I saw it perched on the garden shed watching me. I nearly dropped my compost right there and then."

The Environment Agency continues to insist that there is no cause for alarm. A spokesman reacting to these latest sightings said: "There is no species of bat, or any similar airborne animal, that grows to the size reported. I'm afraid that your witnesses are either mistaken about what they saw, or have been the victims of some sort of prank or hoax."

experts from Torchwood brought in to assess the site have determined that the explosive used was of a type not known. It seems likely from the distribution of the blast and the pattern of fragmentation and damage that the main explosion was detonated in the kitchen area.

Subsequent analysis suggests that the explosive was probably in liquid form and also acted as an accelerant for the ensuing fire. It now seems almost certain that the fissile material was packed into metal drums labelled 'Cooking Oil' and detonated by a concentrated blast of intense heat

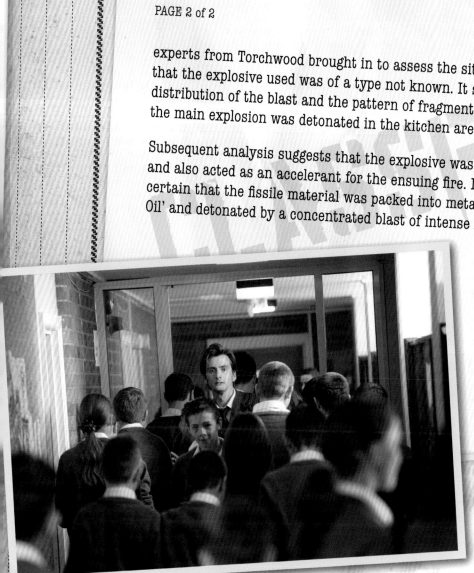

Just what were these children working on?

Humans or giant bats? Or were the Krillitanes originally a different form altogether?

While the 'alien teachers' theory is, as we have seen, wildly improbable, it does explain the strange events at Deffry Vale High School and the surrounding area. Improbable or not, it is interesting to extrapolate from the theory, picking and choosing from other snippets of information presented on the 'Finch-originated' website as well as other sources, including the confidential Torchwood report into the explosion that brought these events to a close.

The nature of the Krillitane creatures themselves is said by some sources to be in flux. They are, according to some, super-evolutionary creatures, able to assimilate traits and aspects of other races into their own bio-genetic make-up. Perhaps this is how they were able to masquerade as human teachers at the school. It is possible that the shape of a huge bat was not their original form at all, but one stolen from another alien race of creatures.

Of the questions that the theory does leave unanswered, possibly the most intriguing is this: What problem or calculation could be so important to warrant such a plan? What were these Krillitanes trying to discover using the enhanced mental capabilities of the children of Deffry Vale?

One suggestion is that they were looking for an elegant solution to Fermat's last Theorem. Another that they wished to learn the secrets of cold fusion. But for sheer audacity, the most extreme theory must be that the Krillitanes were trying to unravel the most basic equations of life itself – that they were within reach of solving the 'Skasas Paradigm', a legendary conundrum which if solved is said to unlock the power of the basic energies of the universe.

Were the Krillitanes playing at God?

What is the blue box? Where did it come from?

Award-winning journalist Sarah Jane Smith — knows more than she's telling?

By its nature, our information about the Krillitanes is largely speculative. What we know for certain amounts to almost nothing – all our knowledge is gleaned from theories that may be completely wrong, and from reading between the lines of official reports. Reports which make it their business to qualify and suppress the truth rather than elucidate.

The one person who might be able to enlighten us further is journalist Sarah Jane Smith, who was actually at Deffry Vale School immediately prior to the explosion. Ostensibly researching an article about the school's impressive exam results and preparing a profile of new head teacher Hector Finch, she has remained reticent about what involvement, if any, she had in the affair. But is it really plausible that a journalist with Ms Smith's track record of reporting on extraordinary and paranormal incidents had no involvement or insight at all?

But if these creatures do exist, then it is clear that they are dangerous, and that they have the facility to appear in human form. They seem to crave positions of responsibility – 'Hector Finch' took on the role of head teacher at Deffry Vale High School. Also, they seek knowledge, almost certainly forbidden knowledge that they would acquire at the cost of other life forms.

If they are indeed super-evolutionary, they may have almost any shape and form of their own, but at Deffry Vale at least they seem to have taken the shape of giant bat-like creatures with the power of flight.

What defence can there be against such creatures? This is unclear – and it is possible there is no defence, except the use of their own Krillitane Oil against them. Somehow, as they continue to change and evolve, perhaps they have developed an allergy to their own oil – which certainly has highly inflammable, not to say incendiary, qualities.

Perhaps, when all is said and done, the best defence might after all be a robot dog.

The best defence against super-evolutionary alien bat creatures?

BEHIND THE SCENES

School Reunion is remembered not only for its distinctive alien enemy, the Krillitanes, and the unnervingly alien performance of Anthony Head as Hector Finch, but also for the return to *Doctor Who* of two favourite 'classic' companions – Sarah Jane Smith and K-9.

Even before the first series of the revived *Doctor Who* had been broadcast, head writer and executive producer Russell T Davies asked writer Toby Whithouse if he would be interested in writing for the programme. Best known at the time as the creator and author of the popular Channel 4 comedy drama *No Angels* – and now renowned for his offbeat supernatural TV series *Being Human* – Toby accepted enthusiastically.

Having renewed his acquaintance with both Sarah and K-9 by watching episodes of the classic series, Toby came up with a story called *Black Ops*. This was set in an army camp and the surrounding village, but after discussions with Russell T Davies, he changed the location, and

```
            Kenny looks in, spooked, hearing…
A snuffle, a crack of bones. From the back of the class.
In plain daylight, something hidden, between the back row
                   of desks and the wall…
    Kenny, still at the front, crouches down, looks…
Right at the back, through all the chair and desk legs,
a black shape. Shuddering. Breathing like an animal. And
then, still at a distance, through all the wooden legs —
FX — a terrible BAT FACE turns, fast, looks at Kenny — !
```

SIGNIFICANT SCREEN ENCOUNTERS

THE TENTH DOCTOR

SCHOOL REUNION
Written by Toby Whithouse
Directed by James Hawes
First Broadcast: 29 April 2006

the story became *School Reunion*.

Since the Krillitanes spend most of the story in human form, it was decided that the actual creatures could be created as computer images, rather than physical costumes. This had been done before, again with creatures that had limited on-screen appearance like the Reapers in *Father's Day*, and the Gelth in *The Unquiet Dead*. Some shots of the Slitheen in *Aliens of London / World War Three* had also used CGI versions of the aliens, so the technology was well-proven.

Effects house The Mill, which handled all the digital and computer effects and graphics for *Doctor Who* at the time, worked with the series' Art Department to design and create the final computer models. The Krillitanes were then added to the live-action material shot in the studio and on location.

BEHIND THE SCENES

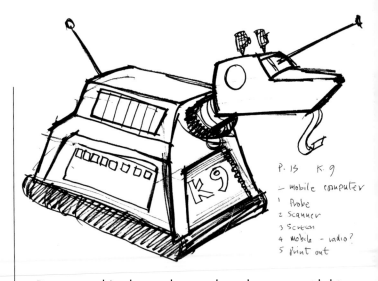

P. 13   K.9
— mobile computer
1 Probe
2 Scanner
3 Screen
4 Mobile - radio?
5 Print out

In addition to the impressive Krillitanes – and Anthony Head's chilling performance as Hector Finch – *School Reunion* is especially memorable for the return of two of the Doctor's best-loved companions: Sarah Jane Smith and K-9.

Sarah, played by Elisabeth Sladen, debuted in the first story of the Third Doctor's final season, *The Time Warrior* (1973–1974). She stayed for the introduction of the Fourth Doctor, not leaving until he was recalled to Gallifrey at the end of *The Hand of Fear* (1976). *School Reunion* saw her return to *Doctor Who* for the first time since a guest appearance in *The Five Doctors* to celebrate the programme's 20th anniversary in 1983.

K-9 also appeared briefly in *The Five Doctors*, now 'owned' by Sarah Jane Smith. This picked up on the storyline of *Doctor Who*'s first ever spin-off – a one-off drama from Christmas 1981 called *K-9 and Company* which saw Sarah teamed up with a new 'Mark III' version of K-9 left for her by the Doctor.

K-9's first appearance was in *The Invisible Enemy* (1977), and he continued to travel in the TARDIS (in two 'versions') until *Warriors' Gate* in 1981. Including K-9 as a regular character was a late decision, with two endings prepared for *The Invisible Enemy* – one where he goes with the Doctor and Leela, and one where he stays with his creator Professor Marius.

Right from the start, K-9 was popular with viewers. And right from the start, the remote-controlled prop caused problems. The first version's remote control systems interfered with the studio cameras, and despite several changes of controls and 'innards', K-9 was never able to travel very easily on any surface other than the flattest studio floor. On location for *The Stones of Blood*, he was pulled along with a fishing line for one sequence, and loaded onto a hidden cart in another – resulting for once in an impressive turn of speed across a field.

School Reunion proved that both Sarah and K-9 were just as popular as ever. It now seems inevitable that they would get their own full series, albeit one focused more on the human of the double act, in *The Sarah Jane Adventures*.

The Nestenes

MASS HYSTERIA, COLLECTIVE HALLUCINATION, even an elaborate student hoax... There have been many explanations for the multiple sightings of shop dummies apparently coming to life. Some of these explanations are plausible, some are ridiculous. None of them is true.

For one thing, there have been not just one but at least two different incidents, decades apart, each involving plastic mannequins smashing their way out of shop windows and shooting down passers-by with guns concealed in their wrists. The official explanation for the waves of deaths that accompanied these incidents is that they were heart attacks brought on by collective hallucination or mass hysteria.

As evidence that this can happen, the authorities point to a similar wave of unexplained deaths across the home counties a year after the first 'mannequin incident'. At the time, these deaths were also the subject of wild speculation about conspiracies or even a mass serial killer on the loose. The so-called 'Flower-Power Killer' apparently left a plastic daffodil at the scene of every death...

But however fanciful and far-fetched that might seem, the answer to these mysteries – and others – may be found in the notes and papers of one of the people who lost their lives in that second mannequin incident in 2005. No one would deny that Clive Finch was a fantasist, a conspiracy theory junkie whose garden shed was a shrine to his own quest to link together bizarre incidents as far diverse as the eruption of Krakatoa and the sinking of the Titanic. But he also documented an organisation known as UNIT. It's a name we've come across before, and there are as many theories about what UNIT is and does as there are uncanny events to which the organisation can be linked.

How much of what Clive Finch 'discovered' could be held up as verifiable truth is debateable. But if his notes and the updates to his website after his death – whoever was responsible for this work – are taken at face value, they imply a terrifying possibility – a collective mind, that can control plastic. A creature that has been colonising other worlds for millions of years, and which has made several attempts to colonise our own world: the Nestenes.

From ersatz Roman legionaries at Stonehenge, to a shower of meteorites falling in Essex; from the plastic daffodils of the 'Flower-Power Killer' to plastic dustbins, from shop window dummies to display duplicates at Madame Tussauds, Finch's theories even suggest the Nestenes could imbue any plastic artefact with inhuman life. A telephone cable, a plastic flower, toy dolls, even plastic furniture could become deadly killers.

But whatever the truth, anyone who has read through Clive Finch's notes and speculation will never look at anything plastic in quite the same way again...

Clive Finch — fantasist or whistle-blower?

Mass hysteria or student hoax? How can these pictures snapped by various witnesses be explained?

149

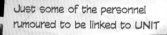
Just some of the personnel rumoured to be linked to UNIT

Alistair Gordon Lethbridge-Stewart — did this man secretly command UNIT?

Quite apart from the papers of Clive Finch, there is plenty of evidence that UNIT exists. Indeed, sections of this book rely on material gleaned from secret UNIT files. But what is far from clear is exactly what UNIT does and who it answers to. When the first rumours emerged of a secret military organisation responsible for dealing with threats that fall outside the remit of other bodies, it was assumed that this 'Intelligence Taskforce' fell under the auspices of the UN. But more recently the UN has been forced to deny any association with UNIT – the latest theory being that the 'UN' in its title stands for 'Unified'.

Certainly, UNIT is an international organisation, drawing personnel from the armed forces of each country where it maintains a presence. Various high-ranking military figures have been associated with UNIT – including Major-Generals Scobie and Rutlidge, Brigadier Winifred Bambera, and Colonel Crichton. Most notably, the late Sir Alistair Lethbridge-Stewart was rumoured to be in command of the UK contingent of UNIT when he was promoted to Brigadier in the 1970s. His memoirs make no mention of the organisation, but the period between 1968 and 1980 is very thinly documented...

So what do we know about UNIT? The answer is very little. But it is possible to piece together some of the events and incidents in which they seem to have been involved. As well as providing security for sensitive projects like the Wenley Moor Nuclear Research Facility, the ill-fated Project Inferno, and the Mars Probe space programme (and the subsequent Deep Space Programme based at Devesham), UNIT were reportedly on hand during the evacuation of central London, working alongside the forces commanded by General Finch (no relation to Clive).

UNIT was also apparently responsible for operations as diverse as investigating (some sources say 'collecting') the meteorites that fell in the area of Oxley Woods, the accident at the Nuton Power Complex, testing the Keller-Machine at Stangmoor Prison (while also overseeing security at the International peace Conference that same month), and, more recently, investigating both the ATMOS debacle and the so-called 'Cube Invasion' – though as we all now know that was an advertising gimmick that got out of hand.

It is instructive that in the whistle-blower leaks of recent years, the one major intelligence organisation not to be mentioned at all is UNIT. Perhaps some secrets are buried much much deeper than others... But for all the rumours and theories, one all-important question remains – whatever its actual responsibilities and powers, is the world safer with UNIT?

AUTO PLASTICS

Wish to report that they are withdrawing their Walky-Talky Doll ('*It Walks, It Talks*') from manufacture with immediate effect.

No orders will be fulfilled.

The Company regrets that it has not been possible to obtain sufficient orders to make the toy viable, despite interest from the Far East.

THE LEGEND OF THE LONE CENTURION

One rare example of a legend that develops and progresses over time is the story of the so-called Lone Centurion. Like the Loch Ness Monster and the Abominable Snowman, the legend depends upon repeated sightings – though in this case of a man rather than a monster.

But this is no ordinary ghost story. The Lone Centurion appears, from all the many accounts, to be solid flesh and blood rather than some ethereal spirit. There are stories – for example the well-known 'Parson's Tale' – of people engaging the Centurion in remarkably learned conversation. Although, in the case of Parson John Tendeter, the Centurion seemed to demonstrate a particularly poor understanding of Latin for a Roman.

The descriptions of the Centurion all seem to tally fairly closely, which is unusual for multiple sightings of what must after all be a mythical character. The first recorded instance dates back to Wiltshire in the second century AD (though there are local stories that may predate this) when the Centurion was sighted standing guard at Stonehenge. Quite why he changed location over the years is a matter of debate, the most plausible explanation being that the Centurion is somehow linked to the 'Pandorica'. Was he originally a Roman soldier charged with guarding the mysterious cube?

This does not, however, fit easily with the more romantic aspects of the legend – that the Centurion is guarding the tomb of his long lost lover, forever hoping that he can revive her. Perhaps, like Sleeping Beauty, the entombed woman could be woken with a kiss…

Those who dismiss the Lone Centurion as rumour and gossip, stories and hoaxes, point to the fact that while there were reported sightings regularly, if not frequently, down the ages, the Centurion has not been seen since 1941. Standing guard outside the National Museum, could he have become a victim of the London Blitz? Or is his task – whatever it was – now completed so that his tortured soul can finally rest in peace…?

From *A Directory of Impossibilities* by Wilson Ryder and Harriet MacMillan, first published 1995.

quite apart from anything else, it seems [...] extreme that a figure such a Cleopatra could have travelled to Britain without there being any corroborative documentation in the main sources. But contemporary historians and Roman writers make no mention of this apparent journey.

So, despite the considerable weight of local 'evidence' to suggest that Cleopatra did indeed visit Stonehenge in this period, taken on balance with the lack of other sources we must put this down to local rumour and gossip. In fact, the most interesting aspect of the affair is that the local population would even have heard of Cleopatra.

While it is tempting to succumb to Professor Jacobsen's theory that the figure hailed as Cleopatra at Stonehenge was in fact someone masquerading as the Queen of the Nile, there is nothing whatsoever to support this suggestion either. As with so much of history, the actual truth of the matter is likely never to be known for sure, and so we can only speculate as to how these stories first started to circulate. But in the final analysis, they must surely be dismissed as a fanciful myth of the same calibre as the Legend of the Lone Centurion — which coincidentally would seem to have its origins in this same period and geographical location…

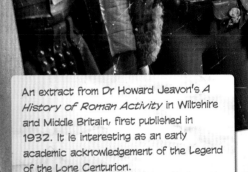

An extract from Dr Howard Jeavon's *A History of Roman Activity* in Wiltshire and Middle Britain, first published in 1932. It is interesting as an early academic acknowledgement of the Legend of the Lone Centurion.

News report from the *Daily Chronicle*. Unusually, there was no follow-up article on the 'man from space' story, although one national paper ran a rather incoherent interview with one Fred Mullins, who claimed to work as a porter at the hospital.

Local Hospital Treats "Man From Space"

According to information received by the Daily Chronicle, it was not just a freak shower of meteorites that recently descended from the skies to land in Oxley Woods.

A source at Ashbridge Cottage Hospital has told our reporter that a "man from space" is being treated in one of the private rooms. Certainly, the hospital seems to have been a centre of attention for the military units drafted in to check the woods for any hazardous materials related to the meteorites. Oxley Woods and the surrounding areas remain cordoned off while the army completes its tests.

The officer apparently in charge of the military aspects of the operation, Brigadier Lethbridge-Stewart, told reporters that the "man from space"

rumours have no basis in fact, although he declined to answer further questions. Dr Henderson, who is said to be treating the spaceman, also declined to comment, although the Head of Pathology at Ashbridge, Dr Lomax did confirm that a specialist had arrived from London. "But we call in specialist medical help for all sorts of ailments," he pointed out. "Not just space sickness!"

HAVE YOU SEEN THIS MAN?

An advertisement that appeared in the *Home Counties Advertiser*. Plastic daffodils were said to be found at the scene of a spate of deaths across the region.

THIS INDIVIDUAL IS BELIEVED TO BE GRAVELY DANGEROUS
DO NOT APPROACH

INFORMATION TO BE GIVEN AT ANY POLICE STATION

This man was sought by British security forces as well as Interpol. It is possible that the original arrest warrant was issued by UNIT. Is it coincidence that the wanted man looks rather like Colonel Masters, who acquired the Farrel Plastics Company? Although, it has been pointed out that the image could equally well resemble the Spanish Ambassador of the time, as well as the newly appointed vicar of the rural village of Devil's End.

Plastic Comes to Town

If you think plastic flowers look artificial and unrealistic, then think again!

Farrel's Plastics invite you to see the results of their revolutionary new moulding process* For a limited time only, we are giving away FREE plastic Daffodils.

Check the list below for the distribution point nearest you.
FARREL'S PLASTIC DAFFODILS – THEY'LL TAKE YOUR BREATH AWAY!
(* Licensed from Auto Plastics)

Following the redundancies reported last issue, it now seems that the ill-fated Farrel's Plastics company is to cease trading with immediate effect.

Following the resignation of owner and manager John Farrel, the company went through a period of modernisation under the leadership of Mr Farrel's son, Rex. However, new lines and automated processes led to rumoured cash-flow problems and the company fell into debt.

It was hoped that the company could be saved when it was acquired by a consortium led by a Colonel Masters. Rex Farrel was kept on as CEO, although Production Manager George McDermott resigned and has apparently left the country. However, despite new lines of products in ranges as diverse as furniture, novelty toys and botanical reproduction, the company was unable to maintain profitability.

The tragedy for the owning Farrel family was compounded by the sudden death of John Farrel from a heart attack last week. Then, in a bizarre accident that is yet to be fully explained, Rex Farrel was apparently shot dead by security forces when he strayed into a restricted area close to the government-run Beacon Hill Research Establishment.

Hopes are fading that a new buyer will be found for the company, although Channing Industries has expressed some interest.

Absence

CAN YOU HELP?

Rose Tyler has been missing from her home on the **Powell Estate** since **6th March 2005**.

Rose is described as 19 years old, 5'4" in height, slim build with shoulder-length blonde hair.

Just one of many people 'missing' after the 5 March 2005 incident — although interestingly, Rose Tyler was reported missing the following day. Her boyfriend was apparently questioned by police, but the matter was eventually resolved when Miss Tyler reappeared a year later.

Anyone with information regarding Rose should contact

...een Rose Tyler ?
...any information ?
...e information line on
...794 6...

Following the release of the government's statement (a part of which is reproduced here), one national newspaper asked readers to send in pictures of the most frightening and unsettling shop dummies they could find. Some of those pictures are reproduced here. Could these mannequins really be dangerous?

Following recent events in London and other large conurbations in Britain, the Home Office wishes to reiterate that there is no cause for panic or alarm.

In addition to the previous statement made by the Prime Minister regarding the unprecedented level of mass hysteria and collective misconception, it now seems likely that public alarm was exacerbated by an ill-judged hoax organised by students less than a month ahead of April Fool's Day. While youthful exuberance and creativity is to be applauded, it does seem that this particular action got out of hand, with some of the participants taking their prank to extremes. The Government urges all those involved to consider the seriousness of the situation and reflect with maturity on the events of last week.

In summary, it now appears that there is no further need for concern. The injuries sustained, damage to property and loss of life sustained during the unrest and subsequent state of emergency are deeply regretted, and Her Majesty's Government again thanks the Emergency Services for their swift and reliable response to the situation.

We can discount the official explanations of mass hysteria and student hi-jinks, since the events described in this chapter have been described by many independent witnesses. In amongst the real hysteria and the attention-seeking hoaxers, an awful lot of information can be gleaned. Add to this the information available to us from UNIT's secret files and reports, and a fuller picture of the creatures we are dealing with begins to emerge.

Shop window dummies and other mannequins

The UNIT files contain two names – Nestenes (sometimes referred to as 'the Nestene Consciousness') and Autons. But, on closer inspection, it seems that the collective entity is known as the Nestenes – Auton being a term used to refer to the animated mannequins manufactured in the first instance by the Auto Plastics company.

Nestenes = Auton

Novelty 'troll' toy

Plastic flowers

But whatever it might call itself, or have been named by the security services, it is clear that this alien entity can imbue plastic – any plastic – with a life of its own. Given our society's dependence on plastic in almost every aspect of everyday life, to say nothing of its military and security applications, this represents a considerable threat.

Just some of the animated – perhaps we should say 'possessed' – plastic artefacts that have been reported are listed below. Obviously, these reports are uncorroborated and should be treated with caution.

Telephone cable

Wheely-bin

RLF 771R

'Waxwork' doubles of prominent military and political figures

Despite the ubiquitous use of plastic, it is important to note that the Nestenes have to deliberately imbue an artefact or object with a portion of their consciousness before it can become active and dangerous. Despite the reports of attempted invasions, plastic remains a safe and harmless substance if used sensibly and correctly.

WEAPONRY

That said, it is vital that we are all able to recognise and respond appropriately to weaponised plastic. In addition to the ways that any normally inert object might try to kill you if suddenly animated with hostile intent, the Autons in particular are armed with their own weapons. These seem to include a powerful blaster-gun concealed in the mannequin's wrist, and revealed when the fingers drop away. Just by pointing at you, the Auton is able to aim its gun.

WARNING SIGNS – HOW TO SPOT AN AUTON

Of course, anyone spotting a shop dummy suddenly coming to life is unlikely to wait around to be shot anyway. But remember that the Nestenes can also animate plastic replicas of real people. You should therefore commit to memory the following tell-tale signs that a colleague, friend, or even spouse may in fact be a Nestene/Auton replica with malicious intent:

- Strange and unusual behaviour
- Inability to remember personal details – although the most sophisticated Auton-Replicas are programmed with a complete brain-scan of the human they have replaced
- Absence of emotion – often betrayed by a lack on inflection when speaking
- Slight 'glossy' appearance of the skin, giving it a plastic-like sheen
- Ability to remould into other forms – for example, hands becoming deadly clubs
- Blaster-gun concealed in the wrist

It would be useful at this point to offer guidance on how best to deal with an Auton. Unfortunately, no information is available, beyond the rather obvious advice: 'Run!'

Could your best friend be made of plastic?

BEHIND THE SCENES

Originally titled 'Facsimile', the *Doctor Who* story *Spearhead from Space* achieved several notable 'firsts'. It was the first *Doctor Who* story to be made in colour, and the first to feature not just the Autons but also the third incarnation of the Doctor played by Jon Pertwee. It was also, due to industrial action at the BBC meaning that studios were unavailable rather than through any advance planning, the first *Doctor Who* story to be shot entirely on film on location. For the entire initial run of the show, the practice was to record the series on videotape, with some filmed inserts from location or film studios where necessary. *Spearhead from Space* was different.

The Autons were different too – a new type of threat for a new Earth-based era for the programme. The fact that they imitated real-world objects – and people – helped to ground the story in a level of

THE THIRD DOCTOR

SPEARHEAD FROM SPACE
4 episodes
Written by Robert Holmes
Directed by Derek Martinus
First Broadcast: 3-24 January 1970

TERROR OF THE AUTONS
4 episodes
Written by Robert Holmes
Directed by Barry Letts
First Broadcast: 2-23 January 1971

THE NINTH DOCTOR

ROSE
Written by Russell T Davies
Directed by Keith Boak
First Broadcast: 26 March 2005

THE ELEVENTH DOCTOR

THE PANDORICA OPENS / THE BIG BANG
Written by Steven Moffat
Directed by Toby Haynes
First Broadcast: 19-26 June 2010

realism rarely attempted before.

The new monsters were an instant hit – the scenes of the Autons smashing their way out of shop windows remain memorable to this day. Though, in fact, no glass was broken – the effect was achieved through clever editing and sound effects. Broken glass or not, the Autons were a smash hit, and returned to battle against the Doctor again in the opening story of the next year – *Terror of the Autons.*

This story explored the Nestenes' affinity with plastic further, introducing a murderous phone cable, suffocating sofa, novelty killer doll, deadly daffodils... It also introduced the recurring character of the Master.

Both stories were written by Robert Holmes – then relatively early in his *Doctor Who* writing career (*Spearhead from Space* was only his third

story for the series). Holmes would go on to write many of the most memorable and effective stories of 'classic' *Doctor Who*, as well as working as script editor for the popular early Fourth Doctor era.

But despite his later successes, the Autons remain one of his most memorable and frightening creations. And they would achieve another notable 'first' in 2005, when Russell T Davies decided to use the Autons as the enemy that would herald the triumphant return of *Doctor Who* to British television.

BEHIND THE SCENES

ROSE — THE NESTENE CONSCIOUSNESS
The Doctor is walking down to the next level, to a platform above the centre of the
vat. The dummies descend, matching his move.
Rose stares, boggles, as a huge blister forms on the surface of the vat. And the
blister stays inflated, settles into the shape of a face — huge, many metres across —
flat on the horizontal, staring up. It's a rudimentary, blank face, blind eyes, like
a basic dummy's features. But the mouth can talk. A voice fills the air. Rose chilled:
harsh alien vowels, reverberating all around.

Despite their popularity, the Autons and Nestenes did not return to *Doctor Who* during its initial run, although there were plans for them to reappear. Until quite late in the pre-production process, the 20th-anniversary special *The Five Doctors* included a sequence where the Third Doctor and Sarah Jane Smith encounter the Autons. In a ruined high street, Sarah is attacked by mannequins from a wrecked shop display, and rescued by the Third Doctor in Bessie.

In 1985, the BBC decided to 'rest' the programme for 18 months. As a result the planned Season 23 was replaced with the 14-episode portmanteau story *The Trial of a Time Lord*. It had been planned for several old enemies to return, with scripts already being prepared for stories involving the Celestial Toymaker and the Ice Warriors. At a less advanced stage of scripting was an Auton story tentatively titled 'Yellow Fever and How to Cure It'. This was being written by Robert Holmes and was to be set in Singapore – where it was planned to go to shoot a large portion of the story.

But despite having missed out on these two classic appearances, the Autons were destined to return – and in style. The Autons – and the controlling Nestene Consciousness – were the first threat to be faced by the Ninth Doctor when *Doctor Who* returned in 2005. More importantly, as Russell T Davies explained to BBC Books at the time, they were the first alien menace to be encountered by Rose Tyler.

'I decided to bring back the Autons – although this time, the Doctor never actually uses their name! – because it was important that Rose, in her first adventure, could consider the whole thing to be one big trick. If, in the first five minutes of the episode, she saw a great big tentacled thing, then we don't have too many of those on Planet Earth – she'd know they were aliens! But plastic, even if it's living plastic, can keep her doubting for a long time, while she gets to know the Doctor. Is it radio control? Clever prosthetics? Smoke and mirrors?

'It was an honour to resurrect the creations of one of the programme's finest writers, the late Robert Holmes – what a

BEHIND THE SCENES

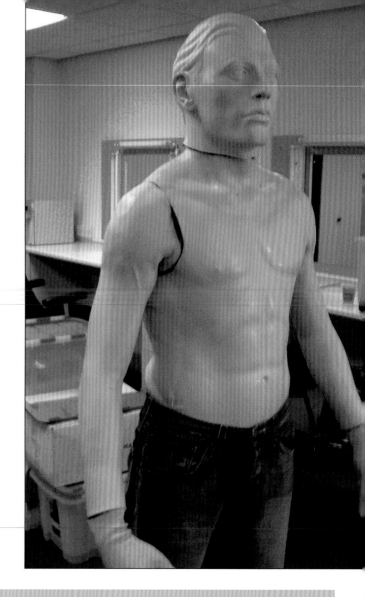

genius! If a small spark of his wonderful mind can touch the new series, then we're very lucky indeed. As a writer, I'd like to think this episode is a tribute to him, with love and gratitude.'

Another reason for using the Autons was that as a threat they are easy to explain. The script could focus on the characters of the Doctor and Rose rather than a complex backstory for the alien menace. Deadly shop window mannequins was all any viewer really needed to understand. Using the Autons also allowed Davies to recreate one of his favourite classic *Doctor Who* scenes where the Autons leave the shops to attack – this time actually breaking the glass windows to step through.

He then did it again, even more effectively, for a flashback sequence in the 2006 story *Love & Monsters*.

The initial script for *Rose* included more Auton action – including a more complicated escape for the Doctor and Rose in the department store as Autons smash their way into the lift, and one

> JACKIE stands frozen. The world has gone insane: DUMMIES firing. SHOPPERS running, screaming. (NB, see the shots fired, but not the actual impact, cut around that.) A MAN has a CHILD DUMMY clinging to his back. CAROLINE and SON run past, screaming. At the far end, a car is burning. Sirens wail. A phalanx of DUMMIES walks down the street. Calm, elegant. JACKIE's crying. Heaving for breath. She's tucked away to one side, backing away from the madness… Behind her: a bridal shop. Jackie's staring out, not seeing the THREE BRIDE DUMMIES in different white dresses step forward, raise arms, swipe —
> - glass shatters, CU Jackie, turning round, screaming —

Auton is sheared off at the waist, and a second Auton Mickey who appears in the finale to trick Rose into revealing the Doctor's anti-plastic plan… These sequences were lost as the script was refined and tightened into what must be one of the most effective 'pilot' television episodes ever written.

The Autons themselves were designed by the *Doctor Who* Art Department, led by Production Designer Edward Thomas. The masks and plastic bodies were created by the Millennium FX company – which has provided prosthetics and monsters for the series since 2005.

Although the Autons returned – as Roman soldiers – in *The Pandorica Opens*, and companion Rory Williams spent time as an Auton in that episode and the following *The Big Bang*, these were 'replica' Autons rather than the blank-faced mannequin versions. ●

The Ood

THE OOD: They came from a distant world. They voyaged across the stars. They sought our humanity. All with one purpose – to serve.

Or so the Ood Operations company founded by the Halpen family would have us believe. But as well as the growing suggestions of problems with Ood production, in addition to the rumours about Ood going mad and turning on their owners, added to the reports of the tragic events on Sanctuary Base 6, other more unsettling stories are starting to emerge.

What if they don't work for humanity out of choice and a sense of purpose at all?

What if it is all a lie?

Let us start our investigation with the facts on which everyone can agree.

The Ood are an intelligent race from the Ood-Sphere, a planet in the same system as the Sense-Sphere in Sector 242.16. Like the Sensorites of that world, the Ood are known to be mildly telepathic. They may discern moods and wishes from this ability, but they communicate with humans by use of technology – a translator ball.

The Halpen family discovered the Ood-Sphere and the Ood, who seemed not only amenable to humanity but wished to serve human beings as a way of giving themselves a purpose in life. And so Ood Operations was set up. The company 'processes' the Ood – fitting them with a translator ball, and telepathic training and induction modules.

A spate of 'special offers' and discounts has given rise to speculation that Ood Operations may be suffering financial difficulties caused by a slowing in the market for their product. The chief executive, Klineman Halpen, has always denied this.

But increasingly, there are stories of Ood going 'rogue' and attacking their human masters – though now, in the 42nd century, we are more enlightened than to believe the Ood could be mere slaves. So what can cause the creatures' eyes to turn blood red, and bring on the aggressive behaviour that has been reported? Are the rumours true – do Ood Operations employ more invasive techniques than they admit to when they 'process' the native Ood?

The organisation Friends of the Ood would certainly have us believe the Ood are exploited. Their latest promotional material suggests that the native Ood have a 'hind brain' removed and replaced with the translator ball. That the Ood are a sort of composite, gestalt race that all share a huge mind embodied in a giant brain that the Halpens found centuries ago. That this brain has itself been suppressed and enslaved. And has recently been set free.

Rumour and gossip? Perhaps. But after the Sanctuary Base incident, can we be sure? Especially as Klineman Halpen himself has apparently gone missing and Ood Operations is said to be about to file for Article 19 Bankruptcy protection.

The true facts behind the loss of Sanctuary Base 6 will probably never be made public – even if they are known to the authorities. What is known is that the base was destroyed, falling into a black hole, with the loss of several of the crew and the full complement of Ood workers.

Funded by the Torchwood Archive, Sanctuary Base 6 was established on the so-called 'Impossible Planet' on the edge of the black hole designated K 37 Gem 5. Why the planet did not succumb to the immense gravitational pull of the black hole has been a scientific mystery for centuries. In the legends of the Veltino, the planet was called Krop Tor – meaning the bitter pill. Legend has it that the black hole was a demon, tricked into swallowing the planet. But the planet was poisonous, and the demon spat it out – hence its name.

Whatever the truth, Sanctuary Base 6 was set up to discover what immense power could keep the planet out of the black hole, and to finds ways of exploiting that power source – which was apparently generating over 90 stats on the Blazon scale.

Right from the start, the expedition was struck by tragedy – the captain was killed in the dangerous approach to the planet, and Zachary Cross Flane took over command.

All that is known for certain of what happened next is that the planet's orbit decayed rapidly and it fell into the black hole. Somehow, the base's escape rocket made it clear, but most of the crew including archaeologist Toby Zed, the Head of Security John Maynard Jefferson, and maintenance trainee Scootori Manista were killed, together with the Ood. Only Flane, Science Officer Ida Scott and Ethics Committee representative Danny Bartock survived.

Did the base's attempts to drill down to the mysterious power source destabilise the planet and cause its destruction? Were all the necessary checks carried out and health and safety procedures adhered to? The official report says they were.

But it is what the report doesn't say that is telling.

Leaked transcripts of broadcasts from the base during those final days raise questions that have never been answered. Who was Rose Tyler? She is not listed on any crew manifest, but it seems both she and an unnamed 'doctor' were present at the end.

What was 'the Beast'? Did the crew come to believe that the legend was somehow true? Were they and the Ood affected by some malevolent force emanating from deep within the planet itself? Is it possible that the Ood were possessed by an alien intelligence that raised their telepathic awareness to over Basic 100, and they rebelled? An intelligence that could only be destroyed by plunging the Impossible Planet – and Sanctuary Base 6 – into the black hole..?

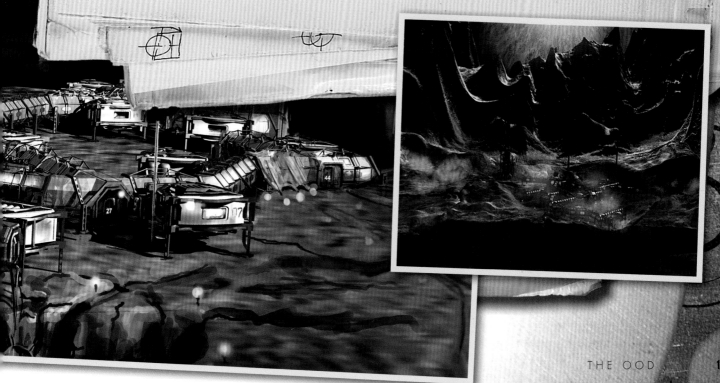

...thus did the disciples of light seek to slay the beast with poison. And the deathless prince, the bringer of despair, did eat of the food that was given as an offering. He tasted not the poison until it was too late. Then he spat out the bitter pill, or so it is given forth, and he did turn his wrath upon the disciples of light.

But the poison worked within the beast, and he did grow weary and lay himself down to sleep. Thus were the disciples of light able to ensnare the beast, the bringer of darkness, and did imprison him for all time in the pit, within the place they called Krop Tor - the bitter pill. Then they devised how they might ensure that the beast and his legions could never again rise and bring night to the universe...

These photographs were recovered from back-up data storage on the Sanctuary Base 6 escape shuttle after the incident.

TRANSCRIPT OF TRANSMISSIONS FROM SANCTUARY BASE 6

This is the final report of Sanctuary Base 6. Officer Tobias Zed, deceased, with honours. 43K two point one. Also Ood 1 Alpha 1, deceased with honours. Ood 1 Alpha 2, deceased with honours…

OOD TELEPATHIC TRANSMISSION 1

These are the words of the Beast and he has woken. He is the heart that beats in the darkness. He is the blood that will never cease. And now he will rise.

OOD TELEPATHIC TRANSMISSION 2

We are the Legion of the Beast. The Legion shall be many, and the Legion shall be few. He has woven himself in the fabric of your life since the dawn of time. Some may call him Abaddon. Some may call him Krop Tor. Some may call him Satan or Lucifer. Or the Bringer of Despair, the Deathless Prince, the Bringer of Night. These are the words that shall set him free:

'I am become manifest. I shall walk in the light and my Legions shall swarm across the worlds. I am the sin and the temptation and the desire. I am the pain and the loss. I have been imprisoned for eternity. But no more. The Pit is open – and I am free!!'

'Artist's impression' of the end of Sanctuary Base 6 — the planet plunging into the Black Hole.

OOD OPERATIONS

Ood Operations Head Quarters

...lphere

Welcome to Ood Operations, serving you since 3914. For over two hundred years Ood Operations have brought you the finest in service options, both in private and commercial capacities, and long may that tradition continue. With new development and quality control systems in place, our service units are seco... to none. When you think service, think Ood Operations.

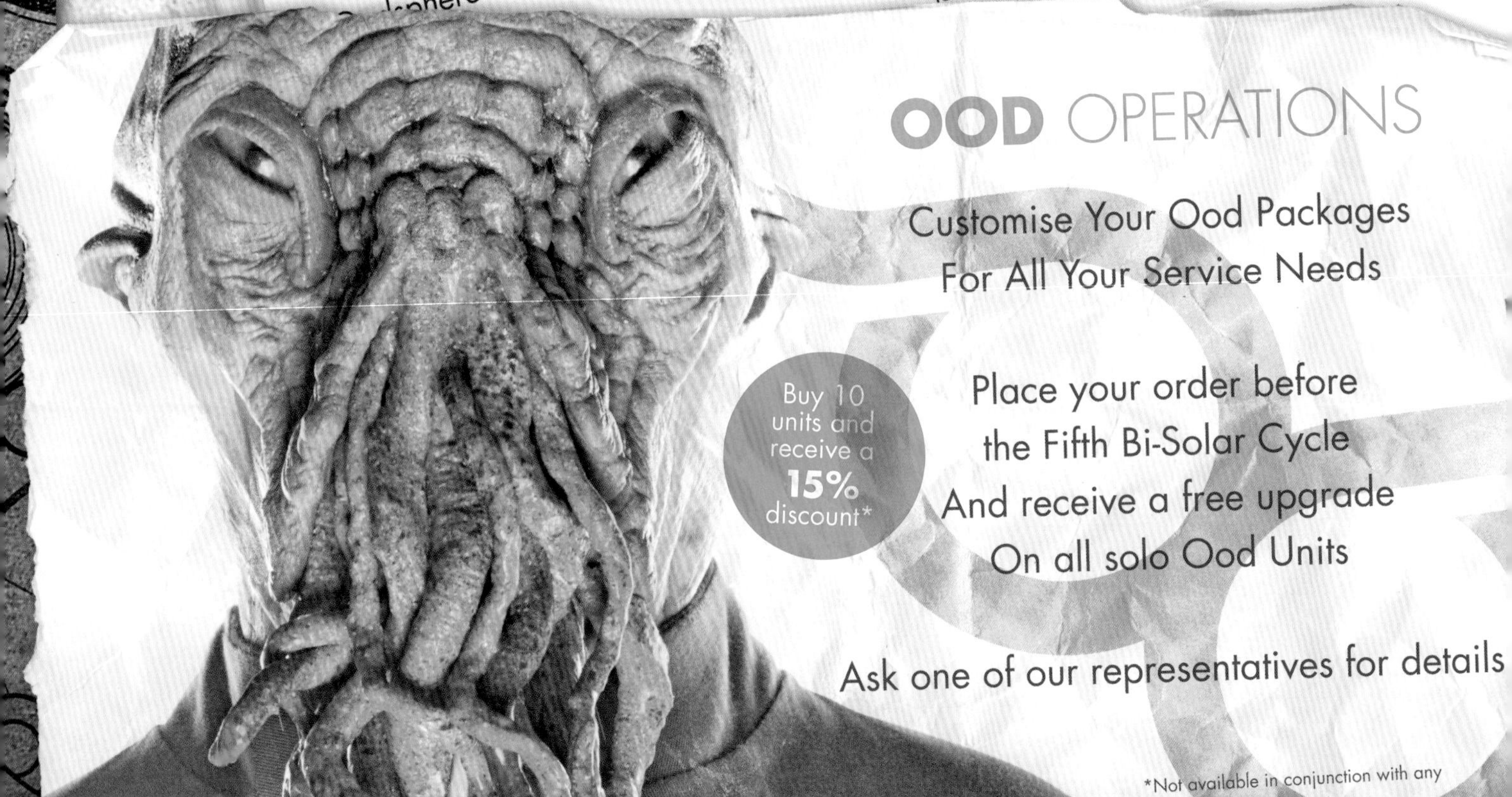

OOD OPERATIONS

Customise Your Ood Packages
For All Your Service Needs

Place your order before
the Fifth Bi-Solar Cycle
And receive a free upgrade
On all solo Ood Units

Buy 10 units and receive a **15%** discount*

Ask one of our representatives for details

*Not available in conjunction with any other offer. Ts & Cs apply.

SERVICE WITH A **SMILE**

The Ood were born to serve, and here at Ood Operations service is everything. We know what you want and what you need, and with over 200 years' experience in the service industry, our reputation is second to none.

Forget robots or artificial intelligence, Ood Operations offer a complete service package in a self-contained, cost efficient and human friendly unit. With an Ood you can go about your daily life, free in the knowledge that all of your service needs are being taken care of.

An Ood for Every Occasion

Here at Ood Operations, we take great pride in being able to provide an Ood unit for every occasion, a product to meet all of your service needs and more. As an intelligent, organic unit, the Ood can be trained and tailored to meet your every need. This makes the Ood, without doubt, the finest service product on the market today. Your wish is their command.

PERSONAL TRAINER **OOD**

The ultimate in service Ood can now custom trained to help bring you to physical peak. Forget those synth supplements with their unpleasa social side effects, hit the heigh physical fitness the old fashio with your very own Person trainer. Fully qualified in species' anatomies and enhancement, your ver Trainer Ood can mak and physically fit ne

Added feature

- Complete a physiologi 67,000

- Physic set y ne

GUIDELINES FOR MAINTAINING A **HAPPY OOD**

he Ood are a self-contained and efficient unit, but some inor unit maintenance is required to ensure that your Ood is ontent and ready to serve. Please ensure that you read the ollowing guidelines for keeping and maintaining your Ood before initiating Ood orientation.

When you receive your Ood delivery pod, your Unit will be in pristine condition and ready to serve. If you are unhappy in any way with your Ood Unit, please contact our customer services department by vidfone (vidfone command - Ood Operations Customer Service) and a sales representative will be happy to assist.

1) On arrival of your Ood Unit, simply greet them with the phrase, "Hello Ood, Welcome to your new home" and a voice recognition chip in the Ood's translator ball will verify your identity and activate the unit so that the Ood may respond.

2) To allow your Ood to acquaint itself with its new surroundings, show the Ood around your living quarters, clearly identifying each area and the Ood Unit's sleeping quarters. Once orientated, your Ood is now ready to serve and receive commands. Your Ood can read and understand your specified language so feel free to give your Ood an e-list of tasks and leave it to carry out these demands if you so wish.

FURTHER INFORMATION

Despite the promises of official reports and the ongoing crusade of Friends of the Ood and other organisations, the truth will probably never be known.

Are the Ood really a servile, 'domesticated' race happy to work for the good of all humanity? Or are they a passive alien race of mild telepaths who have been ruthlessly exploited by an industrial corporation?

If the speculation is true – and it is just speculation – then the Halpens discovered something in the snows of the Ood-Sphere that not only changed their family's fortunes but also affected the development and social existence of an entire race. Where once we were all proud that companies like Ood Operations existed and enjoyed the services of the Ood it provided, most people now accept that the company brought shame on humanity. Its fall from kudos to disrepute has been swift and unmitigated.

And what of Klineman Halpen himself? Has he retired into obscurity as the official statements from the company would have us believe? Or are the Friends of the Ood nearer the mark with their incredible allegation that he was secretly dosed with Ood Graft by his own personal Ood, and has mutated into one of the creatures himself?

Like so many other things – we shall never know.

WARNING SIGNS

But in the meantime, while there are still Ood employed throughout the galaxy, apparently happy in their work, we should be wary. The condition known as 'Red-Eye;' may well exist.

As its name implies, the first warning signs of distraction and lack of concentration in an Ood are swiftly followed by the reddening of its eyes. If your Ood succumbs to this condition, then the official advice is clear:

- Confine the Ood immediately, and if possible immobilise it.
- If it is not possible to confine the Ood, get as far away as you can.
- Hide in a secure place and call the Emergency Services on the usual number, stating: 'Red-Eye incident at [your location]'.
- A security skimmer will be despatched immediately.
- Do not approach a Red-Eye.
- Do not attempt to reason with or control a Red-Eye.

The Supreme Judiciary has sanctioned the use of lethal force in self-defence as a last resort.

BEHIND THE SCENES

For a race of creatures as distinctive and popular as the Ood, it is surprising to learn that they almost didn't appear in *Doctor Who*. One early draft of *The Impossible Planet / The Satan Pit* featured not a new telepathic race of humanoid aliens, but the return of the Slitheen. This was a cost-saving measure but a bit of creative accounting and a frugal design allowed the Ood to replace the Slitheen.

But the constraints of the budget had a direct effect on the design and realisation of the Ood. Executive producer and head writer Russell T Davies showed a picture of a classic series Sensorite to Neill Gorton's team at

THE IMPOSSIBLE PLANET

3 CREATURES are standing there. The OOD; tall, bald, albinos, dressed in simple one-piece suit. White gloves. All more or less identical. Male, but somehow sexless. Eyes white and cloudy, like cataracts. No mouth; instead, painfully red fleshy fronds hang down, like an anemone.

From where the mouth would be, a thin white wire comes out, reaching to a white ball, tennis-ball sized, which each Ood holds out in front, like a weapon. The ball lights up when they talk. Voices unnaturally calm, polite.

BEHIND THE SCENES

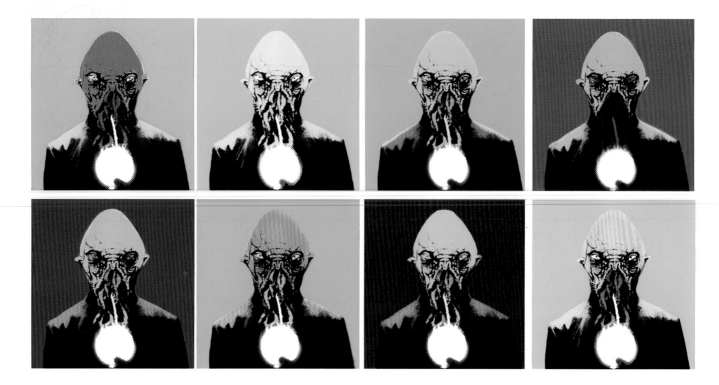

Millennium FX – the company that creates most of the monsters and aliens for *Doctor Who*. Davies wondered if they could produce several essentially identical aliens that could be dressed simply (and cheaply) and given masks cast from the same mould.

Working first from sketches and then a model sculpted in clay, the Millennium FX team decided they could create six Ood creatures. The masks were cast without the 'fronds' hanging from where the mouth should be, which were added later. Modelled on the head of actor Paul Kasey who was to play the leading Ood, the other Ood masks were then modified as necessary to fit the other actors. The masks and fronds were then individually painted.

Kasey's mask was slightly different in one other way too – it was fitted with remote-controlled animatronics to give it some variable expression and blink the eyes.

Despite the tight budget, Millennium FX was finally able to create a total of twelve Ood. Digital effects were used to make it seem like there were many more.

The result was so impressive that the Ood were an instant hit with *Doctor Who*'s viewers. Their return was only a matter of time…

SIGNIFICANT SCREEN ENCOUNTERS

THE TENTH DOCTOR

**THE IMPOSSIBLE PLANET /
THE SATAN PIT**
Written by Matt Jones
Directed by James Strong
First Broadcast: 3–10 June 2006

PLANET OF THE OOD
Written by Keith Temple
Directed by Graeme Harper
First Broadcast: 19 April 2008

THE END OF TIME
Written by Russell T Davies
Directed by Euros Lyn
First Broadcast: 25 December
2009–1 January 2010

THE ELEVENTH DOCTOR

THE DOCTOR'S WIFE
Written by Neil Gaiman
Directed by Richard Clark
First Broadcast: 14 May 2011

```
              PLANET OF THE OOD
         FX SHOT (AND REPEAT THROUGHOUT):
   the gantries overlook a HUGE PULSATING BRAIN.

Reddish-grey, all heaving curls and folds. Like it's
breathing. It fills the entire warehouse, tendrils at
the edges melding it to the walls. And around the walls,
      electric pylons, arcing with blue energy.
```

BEHIND THE SCENES

The Ood-Sphere as seen in *Planet of the Ood* is an icy, snowy planet. To create the effect, the location at Trefil Quarry on the edge of the Brecon Beacons was turned into a snowy wasteland where the TARDIS arrives.

The fake snow was created by the company Snow Business, working closely with Any Effects, the company responsible for providing physical effects for *Doctor Who*. The area was first sprayed with water to wet it down, then a layer of finely-shredded paper was laid down on top of it. The process was relatively quick and before long the grey barren area was turned into a brilliant white landscape.

Ironically, with the shooting taking place in the heat of late August 2007, some of the crew suffered from sunburn as the heat reflected off the brilliant white of the landscape. David Tennant and Catherine Tate wore sunglasses between takes.

Similar treatment was needed in areas of Lafarge's Aberthaw Cement Works in Barry (not far from Cardiff Airport) which was used as the location for Ood Operations. Here the cast and crew had to adhere to stringent safety precautions, except when actually appearing in shot, wearing hard hats, protective boots, and safety glasses. ●

The Silence

HAVE YOU EVER GONE into a room and forgotten what you went in there for? Have you ever lost sight of something you were about to do, as if you'd somehow been distracted? Have you ever wondered where the time went, when something simple seems to take longer than it should? There could be a reason – a straightforward, logical explanation for all these. But it is a terrifying one.

What if the aliens are already here?

What if we see them every day – on the street, in the countryside, at work, even in our homes... We see them, and then as soon as we look away – we *forget*.

Have you ever had the feeling that you've just forgotten something vitally important? Have you ever caught a glimpse of something out of the corner of your eye, only to find that there is nothing there?

But there are clues. People forget, but not everyone forgets completely. We finds hints of the alien presence in art and literature going back for centuries. What we don't know is who the aliens are or why they are here.

Are they simply watching, observing humanity as we make our way through life and develop into what they might deem a sophisticated species? Or do they have a more sinister agenda? Unless and until we can remember them, we will never probably know.

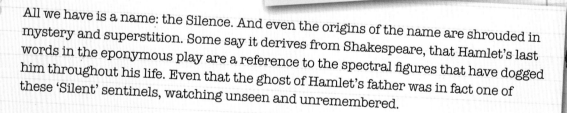

All we have is a name: the Silence. And even the origins of the name are shrouded in mystery and superstition. Some say it derives from Shakespeare, that Hamlet's last words in the eponymous play are a reference to the spectral figures that have dogged him throughout his life. Even that the ghost of Hamlet's father was in fact one of these 'Silent' sentinels, watching unseen and unremembered.

But some academics point out that it is far more likely that the term derives from the fact that while here are hints, clues to their appearance, there is no evidence that the creatures ever speak. Another theory is that they are called 'the Silence' because they are not remembered, because no one ever speaks of them.

Is the Silence a real phenomenon? Or is it a name we have given to that moment of insecurity, that moment when – just for a second – it seems that we've forgotten something important. And that someone is watching us as we struggle to remember...

It is perhaps hardly surprising that over the years a huge number of conspiracy theories have grown up around the theory of the Silence. If we accept that the creatures do actually exist, then many of these theories are plausible. But at least an equal number are, frankly, ridiculous hyperbole of the most sensationalist kind.

One of the most extreme is the suggestion, first raised soon after the event, that the Apollo 11 Moon landing was faked. The theory suggests that it was an elaborate hoax enacted by special effects experts from Hollywood in collaboration with NASA. For years the motive for this was purported to be so that President John F. Kennedy's goal of putting a man on the Moon by the end of the 1960s could be achieved, striking a tangible propaganda blow against the Soviet Union in the Cold War.

But recently another motive has been suggested – that the entire event was faked in order to gain a worldwide audience for the 'live' television broadcast of the landings. This, the theory goes, was because the only way to warn humanity about the Silence would be to do it on a global scale in a single transmission before the creatures were aware what was happening.

As no such warning was given during the momentous broadcast, it is safe to assume that none of the story is at all true. But conspiracy theorists are notoriously persistent, and their position now seems to be that the warning was indeed embedded in the transmission – but that, like the Silence themselves, everyone then forgot about it.

It has also been suggested that a subliminal knowledge of the Silence bleeds through in the work of certain artists and writers who subconsciously channel their forgotten awareness of the Silence into their art. Of course it is easy to read significance into the most insignificant of things. Some of the Art of Silence theories are surely too far-fetched to be credible. But others...

In Pachelbel's famous piece Canon, the music seems to be continually rising – an aural trick in the same way that the steps of Escher's staircase cannot be forever rising as they circumnavigate a waterfall. But is this a metaphor – are both Pachelbel and Escher giving form to the frustration of never quite attaining a goal? Never quite remembering something vitally important?

And is Edvard Munch's famous set of paintings known collectively as The Scream another venting of this frustration? Dating from 1893 to 1910, do the paintings and lithographs show one of the Silence – a figure strangely reminiscent of the 'grey aliens' that have so often been reported? It is surely significant that the 'scream' of the title does not come from the distorted, nightmarish figure depicted in the painting. Perhaps that figure is the reason for it.

~~||||~~ ~~||||~~ ~~||||~~ ~~||||~~

This previously unknown version of Munch's *The Scream* dating from 1902 makes it all the more apparent that the main subject depictured could be one of the mysterious Silence. But who are the two figures seen only vaguely in the background? They also appear in the other versions of the painting.

ATTN: ALL FIELD OFFICES

LEVEL 1 ALERT FOR THE FOLLOWING INDIVIDUALS:

_____ AMELIA POND

_____ RORY WILLIAMS

_____ RIVER SONG (POSSIBLE ALIAS)

ALL ARE ARMED AND SHOULD BE TREATED AS EXTREMELY DANGEROUS.
APPREHEND IMMEDIATELY AND REPORT ANY SIGHTINGS TO DC HQ
PREFIXED BY CODE PHRASE "SILENCE MUST FALL"
LETHAL FORCE MAY BE USED.

A telex sent out to all FBI field offices in 1969. Note the reference to 'Silence must fall', which is a phrase often used in various Silence conspiracy theories. The photographs of the individuals were also sent out to aid identification.

One of the few photographs that exists of what could be a Silent. Experts have cast doubt on this picture as it seems to have been taken with a digital camera of the type used in modern smartphones, yet it dates from 1969.

'Haunted' Home to be Demolished

PERMISSION has finally been given for the former children's home Graystark Hall to be pulled down. The home was closed over three years ago in 1967, after an incident in which a small girl went missing. The child was never found.

Since the closure, stories have persisted that the Hall is haunted. Local kids used to 'dare' each other to sneak in after dark. But the atmosphere in the place was so unsettling that this practice stopped. Kids who did spend time in the building have told this paper that they glimpsed several tall men dressed in dark suits. 'There was something like weird about the face, man,' said one college senior who asked not to be named.

Another student told us that she and her boyfriend had sneaked in to 'make out but nothing heavy, you know', but had been disturbed by the graffiti daubed on the wall. 'It was all like "Get out! Leave this place. Leave now!!" you know?'

The wreckers are due to start work next week, and Mike Hamilton who assessed the building prior to accepting the contract confirmed that there are indeed messages of this sort painted over many of the walls.

People living close to the site tell us that they believe Graystark Hall is haunted by the ghost of a Dr Renfrew – the last director of the home, and the man many blamed for the disappearance of the missing girl. He has not been seen since Graystark Hall closed, and his whereabouts remain unknown.

Note the reference to tall men in dark suits with 'weird' faces in this article from a 1970 edition of the Florida Bugle. It is typical of many possible 'Silence' sightings in that it also involves a haunted building and missing person.

PRESIDENT NIXON
Hello? (Silence) Who is this? (Silence) This is President Nixon, who is calling? ... Is this you again?

CHILD'S VOICE ON PHONE
Mr. President?

PRESIDENT NIXON
This is the President, yes.

This transcript of a telephone call between President Nixon and an unknown child was among the tapes transcribed as part of the US Senate Watergate Committee investigation into the Watergate scandal in 1973.

CHILD'S VOICE ON PHONE
I'm scared Mr. President. I'm scared of the Space Man.

PRESIDENT NIXON
The space man? What space man? Where are you phoning from, where are you right now? (Silence) *Who are you?*

CHILDS VOICE ON PHONE
Jefferson Adams Hamilton.

PRESIDENT NIXON
Jefferson, listen to me...

PHONE HANGS UP.

MALE VOICE (UNIDENTIFIED)
Fellas, the guns, really? I just walked into the highest security office in the United States, parked a big blue box on the rug. You think you can just *shoot* me?

FEMALE VOICE ('MRS ROBINSON'??)
They're *American!*

MALE VOICE (UNIDENTIFIED)
Don't shoot, definitely no shooting!

This transcript, also from the tapes transcribed as part of the US Senate Watergate Committee investigation into the Watergate scandal in 1973, was made in the Oval office in 1969. The 'MALE VOICE' has never been identified beyond the codename he gives, and neither have the man's associates (who are listed as the mysterious 'Male Voice' describes them). Nixon himself refused to name the so-called 'Mystery Man', citing reasons of international security.

MALE VOICE 2 ('THE NOSE'??)
Don't need to shoot us either. Very much not in need of getting shot. See, we've got our hands up. Amy, put your hands up.

PRESIDENT NIXON
Who the hell are you??

CANTON DELAWARE (FBI)
Mr. President, you have to stay back –

PRESIDENT NIXON
But who is he? What is that box?

MALE VOICE (UNIDENTIFIED)
It's a Police Box, can't you read? I'm your new undercover agent, on loan from Scotland Yard. Code name, the Doctor. These are my top operatives – the Legs, the Nose, and Mrs. Robinson.

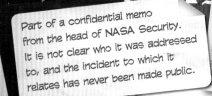

Part of a confidential memo from the head of NASA Security. It is not clear who it was addressed to, and the incident to which it relates has never been made public.

While I can confirm that the unidentified individual was indeed found inside the Command Module of Apollo 11 during the pre-launch period, I am afraid that further details cannot be furnished at this time. As you are aware from my initial report made immediately after we apprehended the man, he claimed to be on a top secret mission for the President. Officer Grant and myself discounted this as the ravings of a criminal seeking to escape justice by any means possible.

However, during our subsequent interrogation of the man, President Nixon did indeed turn up – in person – and vouch for the man, confirming his story.

I appreciate that this is far from a satisfactory conclusion to the incident from NASA's point of view. But equally, we have to accept that the President of the United States is, in effect, our commanding officer and we must respect his desire for secrecy and discretion. I therefore recommend that this incident not be reported to any higher authority, and all records that it ever happened should be removed from our files and security log.

Regards

Chief of Security, NASA

No stars?

Flag moving

Cross hairs

Was the Apollo 11 Moon landing faked as part of a convoluted plan to expose alien activity?

Throughout history there have been stories of shadowy, half-glimpsed, half-remembered figures watching as events and lives play out. Are these stories all derived from the same root? Could they all be more evidence of the existence of the Silence?

The strange, watching figure clothed in black is a staple of ghost stories, of folk tales, of mythology. Tacitus wrote about shadowy figures glimpsed but briefly through a clearing in the fog during his description of the Germanic tribes. Characters appear and disappear without reason in several plays of Shakespeare. Science fiction is littered with references to 'observers' or 'watchers', who use their great powers to observe the universe but are apparently forbidden from interfering in the affairs of others…

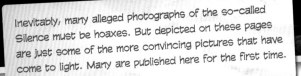

Inevitably, many alleged photographs of the so-called Silence must be hoaxes. But depicted on these pages are just some of the more convincing pictures that have come to light. Many are published here for the first time.

Urban myths and conspiracy theories also seem to draw on the legends of the Silence. What are the mysterious 'Men in Black' if not tall, dark-suited strangers who appear whenever anything out of the ordinary or connected to alien activity takes place. These figures are often said to appear to intimidate witnesses who have seen UFOs or aliens. But perhaps they are not agents of the US – or any other – government at all. Perhaps they come from a race that is well-versed in coercing people to forget what they have seen. Perhaps the Silence are the half-forgotten memory behind not just the Men in Black, but also the archetypal 'Grey' alien…

Do the pictures on these pages actually show these creatures?

Can they really disintegrate a human being merely by pointing a finger?

Are they already here, walking among us unremarked and unremembered even as you read these words?

By their very nature, we shall probably never know for sure.

It is almost impossible to know if you have seen one of the Silence, or witnessed their activity. Often we can recall only the effect and not the cause – see what has happened as a result of the intervention of the Silence, rather than watch it actually happen.

Has a friend of yours inexplicably gone missing? Is there a gap in your memory – perhaps only of a few moments, or perhaps of several days? Is some object no longer where you left it – as if it's been taken away or moved when there was no one there to interfere? Even losing your keys or finding a book on a different table from where you left it could be evidence of the Silence. You have simply forgotten where you put them...

But how can we guard against this? There are a few simple precautions you can take.

First, and most obviously, record your observations. Recording a description or taking a picture on your phone, for example is a possibility. Another is keeping a tally of every time you see one of the Silence. Paper can be lost – or taken – so putting a tally mark on the back of your own hand with a marker pen, for example, may be a more effective option.

There are conflicting views on how useful this can actually be. Some Silence experts insist that the evidence of a sighting also disappears from your memory as soon as you have relived the experience. That would explain why we don't see Silence in every photograph, every video in which they really appear.

But does that mean all the photos are faked or misinterpreted? If not, then we do not yet understand how we can perceive some and not others. Perhaps the Silence themselves somehow decide what information to let through, and what can simply melt away, fading like a blank page.

But one thing is certain. Whatever you do, if you see one of these creatures, it is vitally important that above everything else, before you

Did these people try to warn themselves that they had seen the Silence?

BEHIND THE SCENES

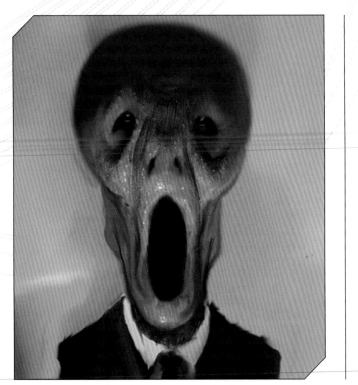

First appearing in *The Impossible Astronaut*, written by *Doctor Who*'s lead writer and executive producer Steven Moffat, the Silence were described in some detail in the script as bulbous-headed creatures with faces like the character in Munch's *The Scream* and wearing dark suits and ties.

Drawing on this description of the Silence, Neill Gorton and the team at Millennium FX created the heads and hands of the creatures, starting with a maquette. This is a small clay figurine that shows what the final design will be like. Once this was agreed, the team could begin creating the full-size masks and gloves for the distinctive four-fingered hands.

Although the Silence are all identical, two different masks were made – one with the mouth

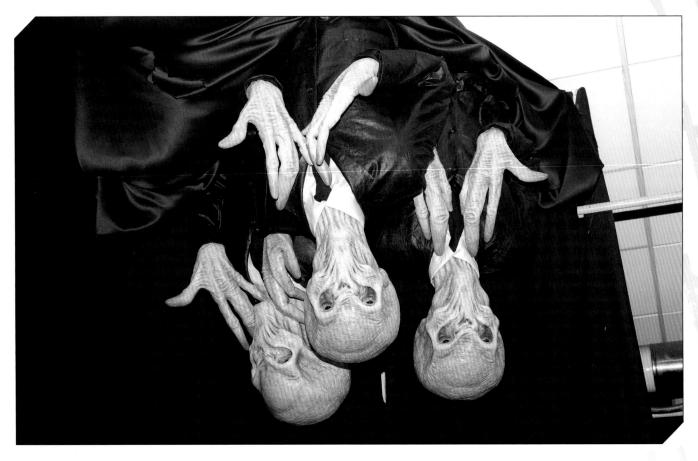

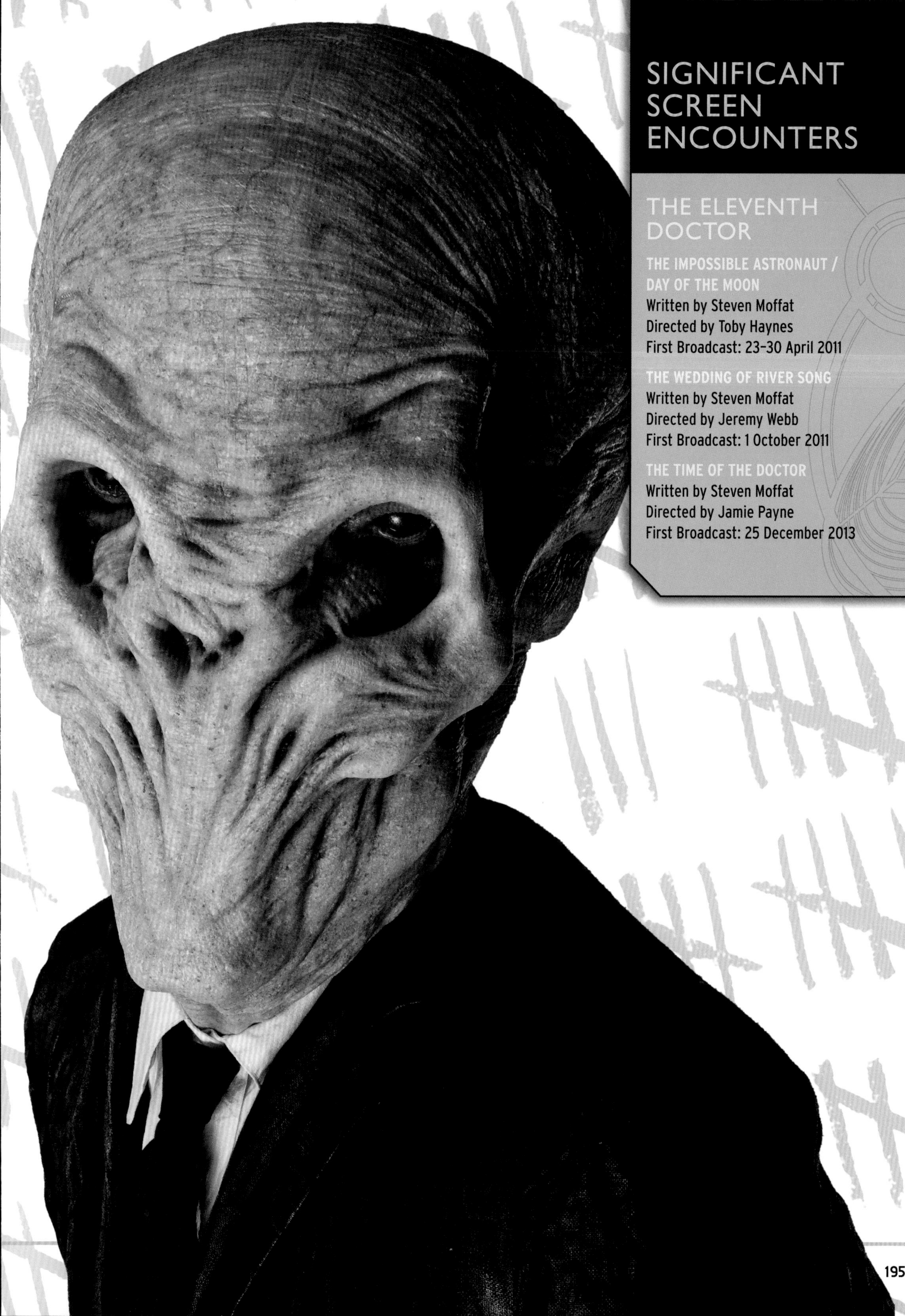

SIGNIFICANT SCREEN ENCOUNTERS

THE ELEVENTH DOCTOR

THE IMPOSSIBLE ASTRONAUT / DAY OF THE MOON
Written by Steven Moffat
Directed by Toby Haynes
First Broadcast: 23–30 April 2011

THE WEDDING OF RIVER SONG
Written by Steven Moffat
Directed by Jeremy Webb
First Broadcast: 1 October 2011

THE TIME OF THE DOCTOR
Written by Steven Moffat
Directed by Jamie Payne
First Broadcast: 25 December 2013

BEHIND THE SCENES

open, and one with it closed. When a Silent opened its mouth, and attacked, the change between the two was actually a computer animation created by effects house The Mill.

The masks were made from foam and latex, and could be inflated with a hand-operated air pump to make it seem like the creature's brain was pulsating.

All the actors cast as Silence – including 'lead' Silent Marnix Van Den Broeke, a Dutch ballet dancer – were well over six feet tall and of very slim build. Even so, the masks were constructed to give even more height, with the eyes of the Silent roughly level with the top of the actor's head. Visibility – through tiny slits lower down the mask – was very limited, and the actors had to be led to their positions on the set for filming.

The suits the creatures wore were made from glazed cotton, so that they looked crumpled and 'waxy'.

The end result was a cross between Munch's figure from *The Scream*, the archetypal 'Grey' alien, and a tall, thin undertaker. ●

THE IMPOSSIBLE ASTRONAUT
Amy, so troubled, is about to reply - when something catches her eye.

Amy's POV: one the Oval Office doors standing open, and standing just outside —

— sudden jarring horror shot!! The BULBOUS-HEADED CREATURE we saw by the lakeside. We're closer this time. From the neck down it's perfectly humanoid, even wearing a simple dark suit and tie, like all the men who work here ...

... but above the collar and tie ...

... it's Munch's The Scream. A swelling balloon head of pink, stretched flesh, those staring dot eyes, that round, aching mouth, that silent scream. And it's just standing there, so still, so calm —

Everyone else watching the Doctor, only Amy seeing this.

On her face, staring: oh my God, oh my God!! She blinks —

FLASH: the image Amy saw on the lake-side, the bulbous-headed creature (which we'll now call a Silent) stepping back into the shadows.

The Silurians

OVER THE YEARS they have been given many names – Silurians, Eocenes, *Homo reptilia*, Lizard people. Even the melodramatic title 'Sea Devils' has been applied to the aquatic versions of these creatures. But whatever we call them, they are a race of ancient reptilians who have been glimpsed across our planet since the dawn of time.

The earliest cited example was the so-called 'Lizard Woman of Paternoster Row'. Whether she was truly some form of lizard is of course, debateable. But certainly there does seem to have been a lady of considerable means who helped Scotland Yard with several of their more intractable and mysterious cases. According to one account, the fabled Lizard Woman was responsible for ending the reign of terror of Jack the Ripper.

Better documented were the Wenley Moor sightings in Derbyshire during the late twentieth century, and the so-called Sea Devil incidents that followed shortly afterwards on the south coast. But these reports, together with the rumours that emerged from the abandonment of the Cwmtaff drilling facility in 2020, have been strenuously denied by successive governments.

It does seem, from comparing accounts of these reptilian creatures, that witnesses have been unable to agree on a description. In Derbyshire, for example, the Silurians were reported as upright humanoids with a ribbed back, clawed hands and feet, leathery faces, deep-set features, and a third eye in the centre of the 'forehead'. In contrast, the Lizard Woman of Victorian London was far closer to *homo sapiens*, albeit green and scaly. She could perhaps have been simply a woman with a severe skin ailment, were it not for the fact that similar descriptions were given of the *Homo reptilia* creatures at Cwmtaff over a century later... The aquatic creatures were different again – with turtle-like heads and large, bulging eyes.

But the suppressed reports of the 2084 massacre at Sea Base 4 are said to include descriptions of both the Silurian and the Sea Devil creatures. The most plausible current theory is all the descriptions are broadly accurate, and that there are colonies of differing but related reptilian creatures secreted across the globe.

How they got there, and where they came from, remains a matter of intense debate.

A selection of images created using the most advanced forensic e-fit technology to show some of the various forms Silurian sightings have taken. (Courtesy of Tea-Lady Forensic Laboratories)

Origins of the Silurians

Gathering information from a variety of sources, some of them classified, it is possible to infer a background for the Silurians. How accurate this history might be, and how much is supposition and extrapolation remains unclear. But the following hypothesis does fit with the primary source in the public domain – the notebooks and journals of Dr John Quinn who was, until his unexplained death, Deputy Director of the Wenley Moor Nuclear Research Facility.

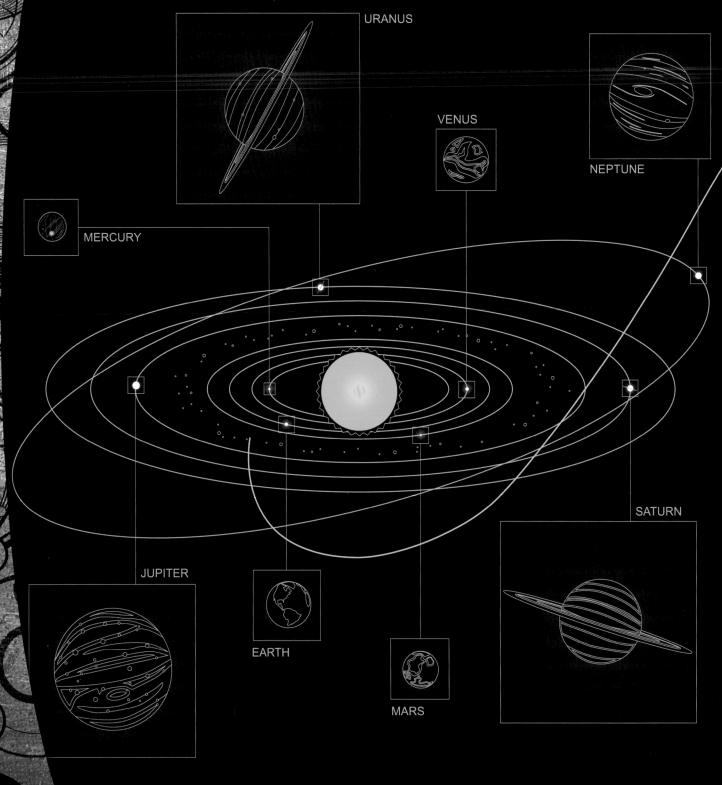

URANUS

VENUS

NEPTUNE

MERCURY

SATURN

JUPITER

EARTH

MARS

A People That Time Forgot

In the dim and distant days of prehistory – some say at the end of the Silurian era, others at the close of the Eocene period, but probably neither is correct – a great catastrophe befell our planet. Another, smaller planet ejected from some distant solar system approached Earth on what seemed like a collision course.

At the time, *Homo sapiens* did not yet exist. Our ancient ancestors were little more than apes, and had no knowledge of the impending crisis. But there was intelligent life on Earth – the reptiles now generically defined as *Homo reptilia*.

The *Homo reptilian* scientists detected the approaching planet, and knew that they had to act. If it did indeed strike the Earth, then all life would end. Or if it came too close, as some of the scientists predicted, it would still cause earthquakes and tidal waves, its gravitational force ripping away the Earth's atmosphere and suffocating all life.

The Silurians – if we can call collectively the various strands of *Homo reptilia* that for the sake of simplicity – devised a two-pronged strategy to survive.

In the little time they had, the Silurians assembled several great Ark ships. On board, they created entire environments – even providing power from the waves of an impossible sea contained in the lower levels. The Silurians collected specimens of various forms of Earth life – plants, vegetation,

dinosaurs… And themselves. Great hibernation chambers were filled with Silurians hoping to escape the catastrophe on Earth and find another planet on which to settle. What actually happened to these ships, no one knows…

On Earth, the same hibernation technology was used in huge underground shelters. If, as seemed increasingly likely, the small planet missed Earth and merely disrupted the surface and temporarily stripped away the atmosphere, then the Silurians could survive, buried deep below the ground.

With the resuscitation systems set to be triggered by the return of the atmosphere, the Silurians knew they would be revived when it was safe to reawaken.

Except, it never happened. The small planet did not steal away the atmosphere. Instead, it was captured by Earth's own stronger gravity and went into orbit around our planet, becoming (at least according to some theories) the Moon. There were earthquakes and horrendous devastation on the surface of the Earth. But as the atmosphere never left, the Silurians' systems never detected its return, and they slept on.

Occasionally, a colony of Silurians is disturbed – by drilling or research or incursion. Occasionally, a group of Silurians wake – and emerge to reclaim the planet they believe they own from the upstart apes that now populate its surface.

THE GREAT DETECTIVE

Is willing to consult with lesser weakling life forms on matters of the mysterious, the unexplained, the remarkable, the enigmatic, and anything else that might be interesting.

Apply in writing or in person to Paternoster Row.

Time-wasters will be obliterated.

Security and strategic attack planning also a speciality – why not let us organise your next conflict?

The Great Detective – leaves other investigators green with envy.

This advertisement, from an 1892 edition of *The Times*, is one of the few tangible items of proof that the 'Great Detective' was not merely a fictional creation but based on truth.

A Study in Green

I confess, I had never before and have never since seen Holmes lost for words. The effect did not endure, but, for the moments that it did persist, I struggled to suppress an uncharitable sense of satisfaction.

Finally, my friend found his voice and addressed the veiled figure who stood before us. 'The mud on your shoes,' he began. Then he frowned. 'No, that cannot be correct. Forgive me, Madame Vastra. The shape of your head, discernible beneath the veil...' His frown deepened.

Her voice was sibilant and soft in reply. 'And you call yourself "the Great Detective".' There was the merest hint of amusement in her tone.

Homes waved away the comment. 'It is an epithet thrust upon me,' he murmured. 'But, please, tell me how myself and Dr Watson may be of assistance to one who shares the soubriquet. Your man Strax was somewhat vague on the nature of the assistance you might require.'

'The large gentleman,' I put in. 'Although the matter to which he referred is now apparent.' I meant, of course, the body lying on the floor between us. 'I formed the impression that Mr Strax is not from around here,' I added. 'He was rather vague about that, too.'

Madame Vastra turned slightly, to look across to where the gentleman in question, her rather solid manservant, was conversing with the maid who had such an appealing smile. They formed a strange pair, and no mistake, the one slim and quick-witted, the other rather more lumpen.

'He is certainly a long way from home,' Madam Vastra mused. 'And I am afraid he can be vague about many things, while remaining infuriatingly specific about others.'

'He mentioned something about fragmentation grenades,' I recalled. 'Is that of some relevance?'

Madame Vastra's veil twitched. 'None at all.'

Her head dipped as she looked down, reminding us of the body that lay at our feet. The blood continued to pool around the battered head. Green blood.

A Study in Green was published only as a limited edition of 200 copies, almost all of which were mysteriously purchased by a young woman dressed as a maid 'on account'. When one bookseller pointed out that the maid did not actually hold an account with them, she apparently replied: 'I mean, on account of how it never should have been published. You want to make something of it?' The bookseller's response is unrecorded.

HIDEOUS LIZARD WOMAN STRIKES AGAIN

UNTIL NOW SIGHTED only in the capital, the repulsive creature known as "the Lizard" has reportedly been glimpsed in the northern industrial town of Sweetville. Unconfirmed reports connect the repulsive reptilian with the sudden and tragic death of local philanthropist Mrs Gillyflower.

TO REMIND READERS, the so-called "Lizard Person" has been reported variously by several witnesses across London. According to some, the creature is female. Others claim it is a villainous knave outfitted in apparel furnished by an unscrupulous theatrical costumier. Possibly the very same theatrical costumier responsible, if that is the word, for perpetrating last Christmas's notorious "Reconstructed Mobile Snowmen" hoax.

WE REITERATE THAT these sightings have not yet been confirmed and our readers should not panic. The local constabulary has so far declined to comment, but a young local resident named Thomas Thomas told us: "She had a troll with her, she did – at the junction of Ambient Way and Methuselah Street, approached by the Great Leeds Road. When they reached Sweetville, I reckon they had arrived at their destination."

An artist's impression of the creature sighted on Wenley Moor. Note that apart from the humanoid, bipedal form it differs significantly from the figure reported in Victorian London.

A short article from the Wenley Gazette.

the 2001-02 season.

...ayed for the side in...

Farmer's Death Remains Unexplained

Police have still made no statement about the incident at Home Farm last week, which led to the tragic death of Farmer Squire. We understand that a post mortem examination returned a finding that the unfortunate man died of heart failure. However, this does not explain the significant police and army presence at the farm following the collapse of Mr Squire.

A source who did not wish to be named told this paper that Mrs Doris Squire made a 999 call to the police shortly before her husband's death. This call apparently related to a possible intruder hiding in the main barn, though Mrs Squire seemed to believe it was an animal rather than a person, possibly some sort of reptile.

Doris Squire remains in hospital suffering from severe shock.

...ecord breaking tee...

Drawing made by Doris Squire while being treated for shock at Wenley Cottage Hospital. She was later transferred to the secure unit at Rampbridge, where she died in 1984.

Day 2

The creature still refuses to speak to me.

Getting stronger. I shall have to be careful. Thermostat still set high. It's like the reptile house at the zoo in here. Not sure how much longer I can keep him locked up. The others will want him back.

But I must know. I must know everything.

How long have these creatures survived down in their shelter in the caves? I've put them down as 'Silurian' but they could be Eocene or even older. I need to speak to a proper palaeontologist, really. This is becoming more than a hobby!

But who can I trust?

Apart from Miss Dawson of course...

IN SUMMARY

That Prisoner 'M' should have been able to make contact with these creatures is clearly an unacceptable breach of security compounded and exacerbated by his subsequent escape from custody.

This committee finds that in addition to displaying a quite astonishing level of naivety, Colonel Trenchard was negligent in the extreme. He broke established protocol on numerous occasions, even allowing Prisoner 'M' to accompany him on an unauthorised visit to HMS Seaspite and supplying the prisoner with electronic components still on the Secret List. Trenchard will therefore receive a posthumous rebuke.

Captain Hart and his team, by contrast, are to be congratulated on handling a situation of extreme hazard in a professional and heroic manner which deserves – and will receive – due recognition. UNIT will of course make its own arrangements for those of its personnel involved, their names and identities not having been released to this committee.

As a final comment before making specific recommendations (Appendix B), the committee wishes to make it clear that this unfortunate situation and the regrettable loss of life would not have arisen if our original recommendation had been followed and Prisoner 'M' either executed or confined to the special security facility at Her Majesty's Prison Stangmoor.

The Royal Navy – Above and Below the Waves

... Perhaps the most interesting insights into the Royal Navy's tactics in anti-submarine warfare come from the local eye witnesses of a naval exercise carried out close to the south coast of England. A frigate and several other warships, including HMS Reclaim, fired depth charges and deployed other underwater weapons – including, according to some accounts, a hovercraft and a '' class submarine carrying experimental SONAR equipment. (As an aside, it is worth noting that Captain Ridgeway does mention the exercise in passing in his memoirs 'Running Silent', but furnishes us with very little detail.) Gunfire and explosions reported at HMS Seaspite suggest there was also a land-based component to the exercise.

Some unintended publicity, not to say amusement, was added to the proceedings by the apparent sighting of large turtle-like creatures said to be washed up ashore and reported on the surface of the sea. A statement from Captain Hart made it clear that these 'novelty plastic masks and costumes' had been accidentally lost from the SS Pevensy Castle, a cargo vessel previously reported missing in the area. Tragically, several ships had been lost in the previous few months, though that can be put down to the prevailing meteorological conditions and adverse currents rather than any connection with the naval exercise, which did not take place until later...

Images taken from data recovered and reconstructed from the CCTV security monitoring system at Cwmtaff.

CONFIDENTIAL

Dear Secretary of State

You will recall that when we agreed to back the Discovery Drilling Project based in Cwmtaff, you put forward a number of potential advantages. I refer you to the minutes of our meeting last year dated 11 October 2019. In the event, I believe that none of these promised advantages have come to fruition. Employment for the former mining area in South Wales was not boosted by the project, since the drilling was almost entirely automated. The more employment-intensive phase of the operation never happened.

As I am sure you remember, myself and other permanent staff cautioned against (I quote) 'another Project Inferno fiasco'. You and your cabinet colleagues ignored these warnings, seduced like Professor Stahlman's backers, I must assume, by the vague promises of a new subterranean energy source.

Now is not the time for recrimination or the apportionment of blame, and I am acutely aware that there has been some loss of life at the project. Following the destruction of the drilling facility, the head of the project, Nasreen Chaudhry, is still missing and chief engineer Tony Mack remains unaccounted for.

Apart from the unexpected (and unhelpful) side effect of turning patches of grass blue, it is difficult to point to any positive benefit to have come from the Discovery Project and my colleagues and I urge you think most carefully before deciding whether to abandon the site entirely or attempt to restart drilling.

I also note that UNIT has passed on its confidential recommendations. While I am not privy to their report, and have no idea what interest the organisation could have in the Cwmtaff incident, I have found from past experience that their recommendations are not to be ignored lightly.

Yours etc

Permanent Secretary,
Department of Energy and the Environment

The Discovery Drilling Project was closed down permanently by the Secretary of State the day after she received this memo from her most senior civil servant.

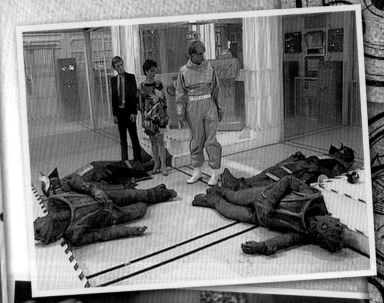

Need an airtight seal…
EVEN UNDERWATER?

Need to be sure that pipe won't leak…
EVEN UNDERWATER?

Need a sealant you can rely on…
EVEN UNDERWATER?

Handy
HEXACHROMITE*
seals it every time

Always test on a small area before full application
Wash hands thoroughly after use. Keep pets away from the affected area.
Non-toxic – though may have an adverse affect on certain types of reptile
and marine life
* Patent applied for

Hexachromite was used as standard for repairs to all Seabase facilities in the late 21st century. Could it have leaked and accounted for the deaths at Sea Base 4?

This fragment of an encrypted video-link message was intercepted by a third party, who deciphered the data and posted it anonymously on the internet.

··· MESSAGE TRANSCRIPT ···

Airlock opening. Now entering Sea Base 4. Keeping our pressure suits and breathing masks on – after 17 days who knows what's happened to the oxygen. You'd think Commander Vorshak would have got out a Mayday at least…

Systems all seem to be working. Evidence of damage though, which confirms what we saw from the outside, particularly Airlock 1… Definitely signs of attack.

<< Break in Transmission >>

We have a body. Repeat, at least one crewmember down. And another, over there…

Air analysis now complete. Getting high readings of… That can't be right – Hexachromite gas?

Proceeding to Bridge. There are bodies everywhere.

<< Break in Transmission >>

Some kind of marine reptile. Huge – but dead. Must be what breached Airlock 1.

<< Break in Transmission >>

Humanoid, at least up to a point. Wearing some sort of armour. Not like anything I've seen before. But there's not much left of the thing. Of any of them…

<< Break in Transmission >>

It's carnage. Everyone is dead – everyone. I have positive identification on Commander Vorshak.

The Sync Systems are burned out. No sign of the Operator, but we know Michaels was already dead…

<< Break in Transmission >>

More of them in here. These seem of a different type, but still very decomposed. Probably affected by the Hexachromite. But what released it? What – or who?

As well as substantial evidence from eyewitnesses and other sources, as we have seen, there is also a body of material that can be gleaned from secondary and indirect sources. Some of this comes from sources that must be treated as potentially unreliable. But much of this material originates within the security services, and in particular UNIT.

Various 'whistleblowers' have provided data and documents that have helped paint a fuller picture of these Silurian creatures. In particular, the investigative journalist James Stevens conducted diligent and comprehensive research into the activities of the UNIT organisation during the period of the Wenley Moor and 'Sea Devil' sightings.

Inevitably much of this material must remain uncorroborated. But it can be extrapolated to make certain deductions about the Silurians. That information is presented here as supposition, and is littered with inconsistencies and contradictions. But it is based on fact.

Third Eye – according to some reports this can be used as a tool or weapon. Others claim it illuminates when a Silurian speaks.

Triad — Silurians seem to organise themselves in groups of three. The Wenley Moor Silurians were led by an experienced Silurian, a younger more 'dynamic' Silurian, and a scientist. The attack on Sea Base 4 was also led by three Silurians.

Another group of three — the so-called 'Paternoster Gang' consisted of the Lizard Woman assisted by a maid and a butler (possibly of Turkish origin)

Hibernation conditions seem to vary. Some 'shelters' are functional and relatively small. Others have been reported to be enormous underground cities.

Among the specimens they preserved were creatures not known today, but adapted for use as weapons by Silurian scientists. One of these was the hideous Myrka — a large reptilian creature that could electrocute its victims.

As well as preserving other life forms on the space Arks, the Silurians apparently took various dinosaur specimens into hibernation.

FURTHER INFORMATION

While there are Silurian colonies and shelters scattered across the whole of our planet, it is important to stress that the chances of you stumbling across one of them are extremely remote. There is a huge amount of subterranean activity across the globe, including leisure activities such as diving, pot-holing and caving. The search for oil and other fossil fuels involves mining and drilling. Archaeologists and palaeontologists dig down deep below the Earth's surface.

Yet very few of these activities have actually disturbed the Silurians. The number of sightings remains relatively small, and each and every one has been accidental, with the single possible exception of the Sea Base 4 incident. Even Project Inferno, the ill-fated plan to drill down below the Earth's crust in search of a new form of energy named Stahlman's Gas, did not encounter any Silurian activity.

But we must not be complacent. If you do discover a Silurian colony, you should evacuate the area immediately. Do your best to disturb nothing, and certainly do not try to explore the facility. The slightest mistake could awaken an entire colony. Report any such discoveries to the UNIT hotline.

Encountering active Silurians, in any of their forms, is also extremely unlikely. Again, should it happen, caution is urged. While the Silurians are intelligent creatures that are able to engage in moral debate and pragmatic argument, it is likely that they will consider you an 'upstart ape'. Particularly if they have only recently woken from hibernation, the Silurians will be of the opinion that planet Earth belongs to them. Your best option is immediate escape. As ever, sightings should be reported to UNIT.

WEAPONRY

If faced with a Silurian incursion, it is important to understand the range of weaponry you may face:

Third Eye – Silurians of the Wenley Moor variety, and possibly others, are able to use the third eye in the top middle of their head as a heat weapon to stun or kill.

Venom – The Silurians encountered at Cwntaff were venomous. Their saliva should be avoided.

Heat Ray – Handheld heat-ray guns were used by the Sea Devils on the south coast and in the Sea Base 4 incident. A variation of the weapon was also apparently used at Cwmtaff. There is no defence against this highly advanced weaponry.

Allies – Other prehistoric creatures are sometimes used by the Silurians to attack people or installations. For example, a dinosaur of unidentified type (possibly Tyrannosaurus) in the caves beneath Wenley Moor, and a deadly electrically charged Myrka against Sea Base 4.

Plague – Unconfirmed reports suggest that the Silurians of Wenley Moor released a biological weapon, possibly a mutation of Bubonic Plague, which resulted in many deaths before UNIT's scientific team developed an antidote.

WARNING SIGNS

It is known that encountering a Silurian can trigger a deep-rooted Race-Memory Response (RMR) in certain susceptible people. The trauma caused by the encounter can render the witness speechless, able to communicate coherently only by means of drawing. These drawings usually take the form of primitive 'cave art' drawn with whatever implement is available on any available surface, even walls. (See for example the drawings made by Doris Squire, reproduced on p.204.) The witness may also be extremely distressed and can become violent. Caution is urged.

BEHIND THE SCENES

The Silurians was the first Doctor Who story written solely by Malcolm Hulke. Previously, he had co-authored two stories for the Second Doctor – The Faceless Ones (written with David Ellis), and The War Games which brought the Second Doctor's era to an end. This ten episode epic was co-written with Doctor Who script editor Terrance Dicks. Already close friends, Dicks was keen for Hulke to write for the series again, but Hulke himself was reluctant. He thought the new format with the Doctor exiled to twentieth-century Earth was limiting. So far as he could see, there were only two types of possible story – revolving round an alien invasion, or a mad scientist.

From discussions with Terrance Dicks, a third

THE SILURIANS

A tiny outlet from the caves. Little more than just a hole in the ground. We hear the sound of a harsh whistling breathing — almost a hissing sound.

We get a brief glimpse of a scaly almost human hand, as the creature pulls itself out of its hole.

possibility evolved – what if the monsters were already here, and pre-dated humanity? Hulke developed this notion into a full story, originally titled simply 'The Monsters' and finally renamed The Silurians (though due to a misunderstanding with the BBC Graphics Department the title shown on screen was actually Doctor Who and the Silurians).

The Silurian costumes were designed by Christine Rawlins from a short description in the script. Since the creatures were prehistoric, she decided their skin should be scaly like a dinosaur's. The head was created by Visual effects Designer Jim Ward, the masks cast from moulds taken from a full-sized clay model of the head. Each included a working light for the third eye, which was worked by a switch the actor operated with his foot. The distinctive voices for the reptilian creatures were provided by actor Peter Halliday (although tests were done with Sheila Grant who had done the Quark voices for *The Dominators*).

For the Sea Devils, the Silurians' aquatic 'cousins', the costumes were completely redesigned as this was intended to be a different species of Silurian. Malcolm Hulke again wrote the script, which was originally titled 'The Sea Silurians'. The Sea Devil heads were created by

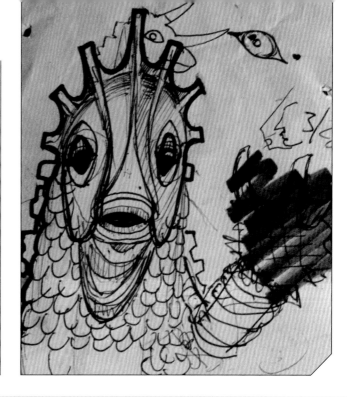

THE SEA DEVILS
(DOCTOR WHO IS COMING ALONG THE CORRIDOR, LOOKING ROUND. AS HE COMES TO AN INTER-
SECTION, HE SEES IN THE SEMI-LIGHT THE SHAPE OF A SEA DEVIL. DOCTOR WHO STOPS.
FOR A MOMENT, DOCTOR WHO AND THE SEA DEVIL LOOK AT EACH OTHER, MOTIONLESS.)

DOCTOR WHO: Don't be afraid. I don't wish to harm you.

(WE SEE THAT THE SEA DEVIL IS CARRYING SOME KIND OF WEAPON. A STRANGELY SHAPED
GUN. THE SEA DEVIL LEVELS THE WEAPON AT THE DOCTOR.)

BEHIND THE SCENES

Visual Effects assistant John Friedlander, who did the initial modelling in clay. They were then worn rather like a hat on top of the head, with the actor looking out through gauze-covered holes in the neck. The design was inspired by the head of a turtle, with large 'gills' added because the creatures were supposed to amphibious. The body of the creature was designed by Maggie Fletcher, and the netting robes were added late on when it was decided the Sea Devils should not be seen 'naked'.

Over ten years later, when *Warriors of the Deep* was made, none of the original Silurian or Sea Devil costumes survived in a fit state to

use (although an adapted Sea Devil mask had popped up in the *Blake's 7* episode *Rescue* in 1981). Again, the monsters were a collaboration between costume designer – Judy Pepperdine – and Visual Effects designer – Mat Irvine.

These new designs were based on but not identical to the originals. The Silurians' third eye now flashed when they spoke rather than operating equipment and tools or being used as a weapon. The original masks for both Silurians and Sea Devils allowed some limited mouth movement. For the Sea Devil leader, animatronics now allowed the mouth to move in time with his speech and blinked the eyes.

It was even longer before the Silurians next returned to *Doctor Who*, and this time they were completely redesigned. Initially, the Millennium FX team led by Neill Gorton that was responsible for creating the new Silurians created a design based on the originals from 1970.

But Chris Chibnall's scripts for *The Hungry Earth* and *Cold Blood* called for far greater interaction between the humans and Silurians and for the Silurians to project a range of moods and emotions. This just would not be possible with a full mask, and so the Silurians were redesigned as prosthetic make-up that would not obscure the actor's own eyes and mouth. This also allowed for individual characteristics to be included in each mask – Eldane could look older,

THE THIRD DOCTOR

THE SILURIANS
7 episodes
Written by Malcolm Hulke
Directed by Timothy Combe
First Broadcast: 31 January–
14 March 1970

THE SEA DEVILS
6 episodes
Written by Malcolm Hulke
Directed by Michael Briant
First Broadcast: 26 February–
1 April 1972

THE FIFTH DOCTOR

WARRIORS OF THE DEEP
4 episodes
Written by Johnny Byrne
Directed by Pennant Roberts
First Broadcast: 5-13 January 1984

THE ELEVENTH DOCTOR

THE HUNGRY EARTH / COLD BLOOD
Written by Chris Chibnall
Directed by Ashley Way
First Broadcast: 22-29 May 2010

A GOOD MAN GOES TO WAR
Written by Steven Moffat
Directed by Peter Hoar
First Broadcast: 4 June 2011

THE SNOWMEN
Written by Steven Moffat
Directed by Saul Metzstein
First Broadcast: 25 December 2012

THE CRIMSON HORROR
Written by Steven Moffat
Directed by Saul Metzstein
First Broadcast: 4 May 2013

THE NAME OF THE DOCTOR
Written by Steven Moffat
Directed by Saul Metzstein
First Broadcast: 18 May 2013

```
                  THE HUNGRY EARTH
And the Doctor leans forward -- and gently, so gently --
  He removes the mask, to reveal our first glimpse of --
     ALAYA: Homo reptilia, warrior class, female.
        A new type of SILURIAN, different tribe.
 Sleek, lithe, sculptured creature.  A skin of green scales,
         but a face of almost human physiognomy.
        And the Doctor marvels at this creature —
```

Restac had a scar down her face…

A disadvantage of the new design was that it took a long time to apply – about three hours for each Silurian. In order to be able to create an army of Silurians on screen, Millennium FX devised protective masks for the military Silurians to wear. These were worn over the face, with the head 'crest' applied to the back of the head but without the need for a full make-up process.

BEHIND THE SCENES

An early draft of *The Hungry Earth* opened with dinosaurs emerging from the ground to attack the drilling project in the pre-titles sequence. In later drafts this was dropped as being too costly and time-consuming to achieve, although writer Chris Chibnall would eventually get his dinosaurs in *Dinosaurs on a Spaceship*, which explores the background of the Silurians' migration from Earth.

But *The Silurians* did feature a dinosaur, similar in appearance to a Tyrannosaurus Rex. This was one of the very first visual effects ever achieved using Colour Separation Overlay (CSO), an early form of what is now known as 'Green Screen', where a key colour in a video sequence is replaced with another video feed – usually to add a background. The dinosaur was a suit designed and built by the BBC's Visual Effects department (and referred to on the script's coversheet as 'Bertram the Friendly

Monster', after the effects assistant who operated him). Only after production was complete did the production team realise they could have made their dinosaur any size using this process – so it could have been a model.

In *Warriors of the Deep*, the Silurians and Sea Devils deployed the Myrka, a hideous amphibious creature that can kill with an electric charge. This was also achieved by means of a suit, this time operated by two performers who wore it in a similar way to a pantomime horse – appropriately enough as the same two actors played the pantomime horse character in the children's supernatural comedy series *Rentaghost*. The results were not as impressive as had been hoped, not least because the studio recording sessions were brought forward to free up studio space ready for the UK's June 1983 general election. •

The Slitheen

FOR SEVERAL DAYS back in 2006, it seemed that the aliens had landed. Or rather, that they had crash-landed. Of course, we now know that the 'spaceship' that crashed into the Clock Tower of the Houses of Parliament was a secret, experimental military aircraft. But for a time even the UK government seemed to believe that the aliens were coming. Who can forget the acting Prime Minister's warning of 'massive weapons of destruction' as he advocated a pre-emptive strike against the invaders?

Then, just as suddenly, it was over. The ignorance and incompetence of Joseph Green, the MP for Hartley Dale – who had found himself thrust into the limelight after the Prime Minister and members of the Cabinet proved 'unavailable' – was defined by that one speech on the steps of 10 Downing Street. It was cemented by his death in a freak gas explosion at Number 10 the very next day.

And with the tragic demise of Green and his colleagues, the media attention quickly switched from the galactic to more domestic politics. The enquiry set up by new Prime Minister Harriet Jones was damning. It detailed the lengths to which certain factions within the previous government and the security forces would go in order to cover up their own mistakes. From the Jones Report we learned the truth of the 'spaceship' that demolished Big Ben, and discovered too that the alien pilot rescued from the Thames (and then apparently shot by the army while trying to escape) was in fact a humble pig, dressed in a spacesuit costume to mislead any journalists who caught a glimpse of it. The truth was out.

Or was it? There are still those who maintain – with some credibility – that the Jones Report itself is a cover-up. That the crash really did involve a spaceship flown by a 'Space Pig', and that aliens managed to infiltrate the very highest levels of Her Majesty's Government – only to be thwarted at the last moment by the intervention of the British armed forces, which fired on their own seat of government.

Could this be true? If it is, then the Jones Report represents one of the most successful and audacious examples of disinformation in modern history.

BAD WOLF

Intelligent alien, or unwitting farmyard animal?

Was Big Ben hit by an experimental military aircraft or attacked by aliens?

One of the creatures sighted in Downing Street — hoax or cover-up? We will probably never know for sure.

How the *Daily Chronicle* suggested the next publicity photograph of the Cabinet might look.

The sudden death of a prime minister almost inevitably gives rise to outlandish theories. The official explanation that the Premier died of a heart attack while on holiday might have seemed more plausible if Downing Street had not been destroyed in an explosion just a day later. The claims by his interim successor that aliens were poised to destroy the world merely added to a volatile mixture of rumour, assumption, and conspiracy theory.

But look below the surface of the wild assertions and official denials, and a frighteningly plausible scenario emerges. What if the Prime Minister was killed as part of an alien plot to infiltrate the very heart of the British Government? His replacement, Joseph Green – a relative unknown in the world of politics – took over at just the time there were rumours of alien visitors, of the assassination of a group of high-ranking military and diplomatic officials at Downing Street. A man propelled into the highest office in the land by circumstances? Or by design?

The most plausible of what have become known collectively as the 'Aliens in Downing Street' theories has it that events were engineered entirely in order to place Green in Downing Street at precisely this moment. That his 'massive weapons of destruction' speech and his representations to the UN and NATO were all pre-planned, were all part of some complex alien strategy that depended on the UK launching its nuclear deterrent.

The evidence may be largely circumstantial, but it is there. Taken individually, the various events, the apparent mobile phone pictures of aliens, the alleged sightings of large green creatures bursting out of human disguises, all seem incredible and implausible. But taken together, a coherent story begins to emerge.

The final piece in this jigsaw of conspiracy is the destruction of 10 Downing Street. Eyewitnesses across London reported seeing a small aircraft or rocket passing over the capital at high speed moments before the explosion. It may have been discounted by the Jones Report, but we should not ignore the contention by retired Commodore Henry Hanrattigan that what he saw heading for Whitehall that day was a Cruise missile.

A frightening scenario begins to emerge. What if Downing Street was indeed infiltrated by aliens who planned to turn the UK's armaments against humanity? As a final option, as a last resort, wouldn't the British armed forces take decisive and extreme action? Wouldn't they launch a surgical strike on the aliens – even if it meant destroying their own Prime Minister's residence?

Assuming this theory is true, and it certainly explains the majority of the evidence, the only real questions remaining are: Who were these aliens, and what did they want?

These pictures of the so-called Big Ben Crash were taken by avionics expert Michael Tucker and used to support his theory that the crash was engineered. Says Tucker: 'The angle and flight-path are just so perfect — it's like the pilot was trying to hit the tower.'

Was Rose Tyler brought back by spacemen? In fact, she returned home before the spaceship collided with Big Ben. This is just another example of news media reporting illogical and unsubstantiated rumour as possible fact.

NEWS LONDON

Home   World   UK   **England**   N. Ireland   Scotland

Space Traveller's Return?

Police and family alike confess themselves baffled by the sudden, unexpected return of Miss Rose Tyler to London last week. Missing for a year, Miss Tyler arrived home to greet her mother, Mrs Jackie Tyler, as if nothing had happened.

Friends say that Miss Tyler claims to have been "travelling", and her return was certainly a relief to local mechanic Mickey Smith, Miss Tyler's former boyfriend. He was arrested on suspicion of her murder soon after she disappeared last year, and has always maintained that Miss Tyler had simply "gone away".

But where did she go? She sent no postcards, no emails – had no contact at all with her family. And is it just coincidence that she arrived home at the same time as a spaceship crashed into the Houses of Parliament? Miss Tyler was unavailable to comment, but some local people speculate that she may have been abducted by aliens, and returned in the very craft that collided with Big Ben in such spectacular fashion. This would certainly help to explain why she seems to have been spotted, along with an unidentified man, entering 10 Downing Street soon after the crash.

BBC

To: RM

From: D.Prog

Given the media attention, not least from our own channels, and the inevitable speculation about a possible alien threat, together with the PM's comments about 'massive weapons of destruction' and the worsening international situation in the wake of the Big Ben Incident — I wonder if it was appropriate or sensible for this week's "Blue Peter" to promote the creation of "space confectionary" for children? Baking a cake might also be seen as pandering to the incoming Prime Minister, given his former position as Chairman of the Parliamentary Commission on the Monitoring of Sugar Standards in Exported Confectionery.

I would be grateful for your thoughts on this, together with any suggestions you might have for mitigating potential adverse feedback.

An actual alien 'Slickeen', or an unconvincing fake?

THE DEATH OF DEMOCRACY

Stunning pictures from Downing Street yesterday – when a suspected gas explosion ripped through the very heart of British Government. The clear-up continues as Harriet Jones calls for an Official Enquiry.

BAD WOLF

224

While there is little tangible evidence to support the more outlandish of these alien infiltration theories, some information does seem more plausible and has been corroborated from multiple independent sources. Perhaps the least contentious is the name of the alien creatures – who, according to several investigators, call themselves Slitheen. There is a suggestion that this may not be a generic name, but rather the name of an organisation, or possibly a 'family' of aliens.

It also seems probable that these Slitheen are a calcium-based life form. It may seem strange to imagine creatures made of the same material as limestone, but life on Earth is of course based on carbon. Human beings are fashioned from the same raw materials as coal and diamonds.

Assuming the calcium connection is accurate, then this also provides us with a possible defence against the creatures. Strangely, it is in antiquity that we find a possible weapon to use again the potential invaders. It is said that Hannibal crossed the Alps by using vinegar to dissolve boulders that were in his way. It is possible that the Slitheen aliens, their structure already weakened by use of the compression field in their body suits, could be fatally injured by the deployment of vinegar and other acid-based food products.

By all accounts, the Slitheen are a technologically advanced, calcium-based species that have evolved as hunters with a highly developed sense of smell. Is it really possible that they could be defeated with a jar of pickled onions?

BAD WOLF

The suggestion that aliens could infiltrate the heart of British government and disguise themselves as senior politicians is a frightening one. But the implications stretch much further. These are not just aliens who can imitate human beings – they are aliens who can imitate specific human beings, and do it so effectively that even close friends and colleagues would seem to be unable to tell the difference.

The latest thinking is that such a thorough disguise could only be achieved by use of synthetic body suits modelled on the actual individuals involved. This implies a high degree of planning in the process. But it seems there are side effects.

If, as appears to be the case, the aliens are larger than the humans they are imitating, then a compression field also needs to be included in the body suit. In this case, excess energy would build up within the suit and could be highly visible – possibly as electrical discharges – when the suit was opened.

Is this the 'skin' of an alien infiltrator?

What horror might lurk inside your flatulent friend?

People with 'fuller' figures would make more natural targets for alien imposters.

Perhaps more obviously, any natural gas build-up within the suit, exacerbated by the compression of the alien inside, would periodically vent to the outside world. This gas release would resemble the natural bodily function of flatulence. If the theory that these aliens are a calcium-based life-form is correct, this alien 'farting' (for want of a better term) would be accompanied by an unpleasant and persistent odour.

It may seem bizarre, but the way to tell if your best friend is actually an alien imposter might well be to keep track of how often, and how impressively, they break wind.

BEHIND THE SCENES

Although *Aliens of London* and *World War Three* were the fourth and fifth episodes to be transmitted in the new series of *Doctor Who* when it returned to television in 2005, they were actually shot first, along with the opening episode, *Rose*. While *Rose* featured the Autons, the Slitheen became the first major new monsters designed for the returning series.

The starting point for the design was of course the description in the script by writer and executive producer Russell T Davies. From this, the designers, led by production designer Edward Thomas and concept artist Bryan Hitch, created initial sketches and drawings. Dan Walker and Neill Gorton of Millennium FX then created the final creatures.

The costumes were constructed largely from latex, and were extremely bulky. They were also very heavy and uncomfortably hot for the actors to wear – especially as recording took place during a period of particularly hot weather.

SIGNIFICANT SCREEN ENCOUNTERS

THE NINTH DOCTOR

THE NINTH DOCTOR
Aliens of London / World War Three
Written by Russell T Davies
Directed by Keith Boak
First Broadcast: 16-23 April 2005

BOOM TOWN
Written by Russell T Davies
Directed by Joe Ahearne
First Broadcast: 4 June 2005

It's eight feet tall, a thick tube of solid, wet, green flesh, all bristling with spikes and spines. The whole thing curves over at the top, like an upright prawn, so its head leers down. A face like a big, sweet, bloated green baby, with jet-black eyes. Green slime trickling from its terrible smile.

BEHIND THE SCENES

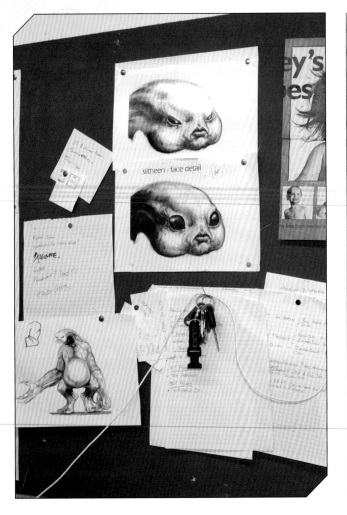

But for some sequences, having an actor in a Slitheen costume proved infeasible. The suits were not designed for rapid movement, and the heads 'waggled' if the actor inside tried to run. This meant that the various chases through 10 Downing Street had to be carefully planned.

Sometimes clever editing and rapid intercutting made it appear that Slitheen were hurrying through corridors when in fact they were hardly moving. On other occasions, computer-generated Slitheen were added to 'clean' images of the background and animated to show the creatures moving at speed.

Another sequence that was complicated to realise was the Slitheen shedding of their human guise. Rather than try to have an actor in a Slitheen costume inside a human costume, the actor wore an all-over green body suit (marked with black crosses to give reference points to the animators). The actors inside the 'human suits' could then be replaced in post-production with computer-generated Slitheen apparently squeezing out of their human bodies.

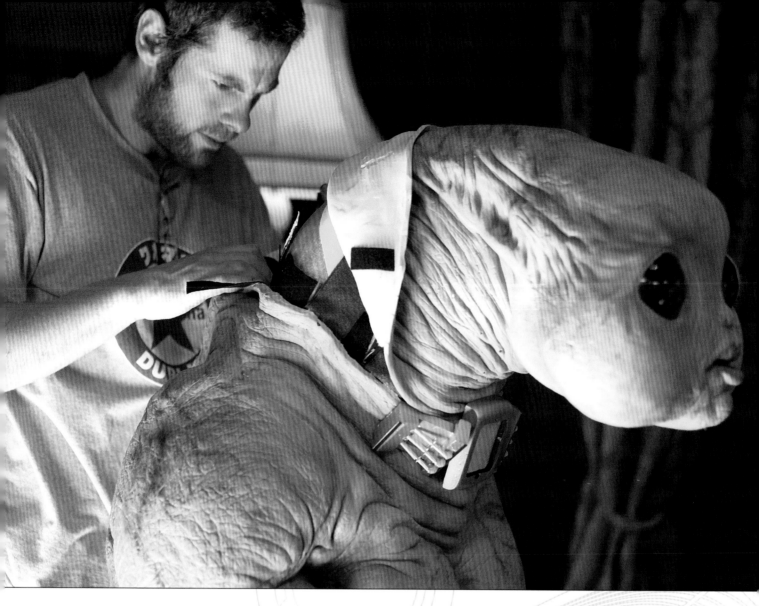

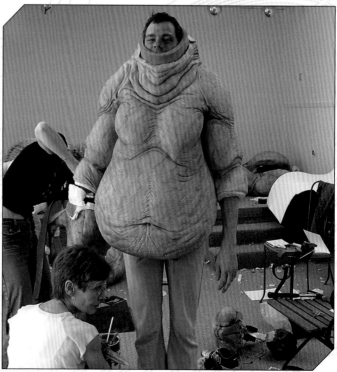

BEHIND THE SCENES

Like the Slitheen themselves, the sequence where the spaceship crashes into the Clock Tower of the Palace of Westminster was achieved using a combination of Computer-Generated Images, and live action.

The spaceship, designed by concept artist Bryan Hitch, was realised by digital effects company The Mill. The computer-generated ship was added to footage of central London as it flew towards the Clock Tower. But the Clock Tower itself, and the wing of the spaceship that slices into it, were real models constructed by former BBC Visual Effects Designer Mike Tucker and his team at the Model Unit.

The model of the Clock Tower was about 20 inches across and took four weeks to build. To achieve the effect, the model spaceship wing was swung through the model, destroying it – except that on the first take, the clock face fell out in one piece rather than shattering as intended. Luckily – and with some prudent planning – the model had been built so that the breakable section could be replaced for a second shot. This time it worked perfectly. Shooting at high speed then replaying it much more slowly gave the sequence a proper sense of scale.

The spaceship's unfortunate occupant – the adapted farmyard pig – was created by Millennium FX. The costume was worn by actor Jimmy Vee. The scenes shot at Albion Hospital (actually Cardiff Royal Infirmary) were the first live-action sequences to be shot of the new series of *Doctor Who*, starting on Sunday 18 July 2004. ●

The Sontarans

THE MARKET TOWN of Ilchester lies in the heart of the ancient kingdom of Wessex. A few miles outside the town, at the edge of a small wood called Sir Edward's Copse, is the inauspicious site of one of the most intractable archaeological mysteries of recent years. In common with many archaeological sites there is almost nothing to see – just fragments of stone and rock scattered across the landscape. All that is left of a once mighty castle.

Records from the early medieval period are scarce. The castle doesn't even have a name, being referred to in the documents that do survive merely as 'Irongron's Castle', taking its name from the robber-baron who was the castle's last owner. But it isn't the ancient records, or lack of them, that excited the historians and archaeologists when the site was surveyed in the late 1980s. It is the modern evidence of what happened to the castle.

The dig, led by Philip Bixby, was in effect a training exercise for his students. He didn't expect to find anything very interesting. Perhaps he wouldn't have done were it not for the fact that one of the team was a mature student named Geoff Carney who had previously served in the Royal Engineers. Looking at the pattern of the scattered stone and examining some of the larger fragments, Carney came up with a remarkable theory: all the evidence suggested that Irongron's Castle had been destroyed in a huge explosion.

This was certainly possible, if not very likely, as gunpowder was manufactured in Britain as far back as the mid-fourteenth century. The theory prompted Bixby to extend his trenches and dig deeper. What he found next was truly astonishing – the remains of a breach-loading rifle. Given the position of the find, beneath the remains of a castle wall, Bixby contended that the rifle must have been inside the castle when it was destroyed – centuries before such a weapon was known to exist.

Further anachronistic finds led experts, including Bixby, to conclude that the artefacts had been planted, perhaps to discredit Bixby himself. Machined metal, complex modern circuitry, even a crude positronic control system were all excavated at the site. The more bizarre the finds, the more obvious it became that they had nothing to do with the historical context. Today, the incident is largely forgotten...

But what if the artefacts were not in fact placed there more recently? What if they really do date back to the early medieval period? Among the documents from the time that do exist from the archives of Sir Edward of Wessex, another local aristocrat, there is mention of 'Irongron's Troll'. The story is that Irongron was feared and hated not only because he was ruthless and cruel, but also because he had supernatural help – in the form of this 'Troll'.

The sources disagree about the origins of the Troll. Some say he was a Saracen, 'skilled in Eastern Magick'. But others claim that he fell from the sky inside a star and that he agreed to help Irongron and provide him with 'star weapons' in return for Irongron assisting him to return to a great battle being fought in the heavens.

Did Irongron give refuge to an alien who crashed to Earth, who in turn provided him with advanced weapons including breach-loading rifles? And in entering into this pact, did Irongron somehow bring about the destruction of his own castle?

L—R: Unidentified woman; unidentified man; Lord Palmerdale; Colonel Skinsale; Adelaide Lessage; Vince Hawkins (junior lamp-man)

It is perhaps not surprising that Irongron's Troll is mentioned in the leaked UNIT files. What is surprising is the context. What links a medieval robber-baron with an Edwardian lighthouse, a murder in a restaurant in 1980s Seville, and the ATMOS Incident of 2009?

Fang Rock is a small, craggy island just off the English coast. Being close to major shipping lanes, there has been a lighthouse on the island since the early nineteenth century. Despite being otherwise uninhabited, Fang Rock has always had a reputation for the uncanny. Locals talk of the Beast of Fang Rock, which apparently attacked and killed several people in the area in the nineteenth century. But the incident referred to in the UNIT files took place in the early twentieth century.

Details are scant, but it seems that a steam yacht was wrecked on the rocks during an especially thick fog. There were several fatalities in the crash, including the ship's captain, but survivors included Lord Palmerdale and his secretary Miss Lessage, along with Colonel James Skinsale the Member of Parliament for Hartley Magna who was a prominent if junior member of the government at the time.

Once the fog had lifted, several vessels reported that the Fang Rock lighthouse was 'dark', and a boat was sent from the mainland to investigate. The bodies of Palmerdale, Skinsale, and Miss Lessage, together with an unnamed sailor from the steam yacht and the three lighthouse keepers were all discovered either inside the lighthouse or on the rocks close by.

The Fang Rock lamp gallery

Madame Vastra's 'man'servant Mr Strax

The incident has never been explained. Given the date, some academics suggest that it was this incident and not the earlier unexplained disappearances on the 'Flannen Isles' in December 1900 which inspired Wilfred Wilson Gibson's famous 1912 poem Flannen Isle. Does UNIT believe that the alien race of which Irongron's Troll was a member was to blame for the deaths?

The 1985 death of popular English bon vivant Oscar Botcherby at Las Cadenas, the restaurant he managed in Seville, is harder to link to alien intervention. The Spanish police report is clear that Botcherby was stabbed following an argument with a customer who refused to pay his bill. The murderer was never traced.

More intriguing is the suggestion made by some theorists that Mr Strax, the burly manservant of Madame Vastra, the so-called Victorian Great Detective of Paternoster Row, might have been of the same race as Irongron's Troll (see p198). This theory has yet to be proven.

Blueprints for the ATMOS device

Luke Rattigan — inventor of ATMOS

A more pertinent link would seem to be the ATMOS Incident. It is clear that UNIT was involved, and there has long been speculation that the incident was linked to an attempted alien invasion. ATMOS – the Atmospheric Omission System – was developed by teenage genius Luke Rattigan and at the time of the 2009 incident was fitted to approximately half of all petrol and diesel vehicles in the world. The system's main function was to strip all carbon dioxide from engine emissions, although it also provided satellite navigation and smart vehicle-management functions.

(Left) Colonel Mace led the UNIT raid on ATMOS. (Right) Captain Price of UNIT, also involved in the ATMOS raid

But something went wrong with the system, and for an extended period vehicles fitted with ATMOS pumped fumes out into the atmosphere. People were warned to stay indoors, and there was a mass migration from denser centres of population to more rural areas until the fumes were burned off by freak solar activity. The official explanation was that the ATMOS units worked by storing up and compressing noxious gases from the engines. Once the system was saturated, the fumes were expelled.

ATMOS was manufactured at seventeen factories around the world, the main facility and distribution centre being based in London. Even before the ATMOS Incident, UNIT are reported to have seized control of this facility, as well as despatching personnel to the Rattigan Academy, situated close by. Interviewed by the media following the incident, local residents seemed to have no inkling of any suspicious activity at the facility prior to the raid. One point that was raised in the local press was that none of the workers at the factory seemed to be recruited from the immediate area, which was a cause for concern.

Luke Rattigan – the only person who really understood ATMOS – disappeared following the incident, leading to inevitable speculation that the invention was a scam. Certainly, he became extremely rich from ATMOS, as well as from his Fountain Six internet search engine.

But if the UNIT files are to be believed, it was not Rattigan who developed ATMOS at all. The files suggest that the system was a 'Trojan Horse' device designed by aliens to contaminate Earth's atmosphere and convert the planet into a 'Clone World'. From their description, these aliens could indeed be the same species as gave rise to the stories of Irongron's Troll. According to UNIT, they are called Sontarans.

None of the ATMOS workers were recruited locally.

An artist's depiction of Irongron's Troll, derived from contemporary descriptions. Was the 'troll' actually a Sontaran?

Part of a manuscript held in the library of Edward of Wessex. It is not clear whether it is supposed to be an accurate description of a battle, or a fiction devised to reflect well on Sir Edward. Note that in this account, Edward has been given his own 'Wizard' to counter Irongron's Troll, almost certainly a fictional addition whatever the veracity of the rest of the account.

The King had levied tribute and so Sir Edward and my Lady Eleanor had but few scant men at arms to defend the castle, together with young Hal the Archer who was much skilled with bow and arrow. But at that time there did come to Wessex a mighty Wizard and his Familiar, a fair maiden full of fire and bravery. Together they did fashion enchantments the like of which we had not seen.

And with this wizardry did Edward and his few defend the castle gainst the might of Irongron, and his troll which men call Toadface. With false armies fashioned from cloth and wood did the Wizard defend the battlements, and created such a stench as Hell itself might breathe out across the land. So was Irongron vanquished and that very night his castle was laid waste, and both he and the troll with it.

Horror at Fang Rock

Fang Rock has long been associated with disaster. The very name of the small island evokes images of the sharp-toothed rocks that have ended many a vessel's final voyage. Happily, since the commissioning of the Fang Rock Lighthouse in 1821, shipwrecks have become rare. With the advent of electrical power to provide an even brighter light, it was supposed that Fang Rock would become even safer.

Alas, this has not proved to be the case. Tragedy has again struck Fang Rock, during the unaccountably dense fog that descended on the area last week. After the fog dispersed, several vessels reported that Fang Rock Lighthouse was unlit, and the local coastguard in concert with the constabulary despatched several officers to investigate.

Upon arrival, the officers found a scene of devastation. As well as extensive damage to the exterior of the lighthouse, they discovered several dead bodies. One of the lamp keepers was so mutilated as to be almost unrecognisable, while the other two showed no signs of injury. Also discovered were the bodies of Lord Palmerdale and his secretary Miss Adelaide Lessage, as well as that of Colonel James Skinsale MP. More corpses were recovered from the wrecked remains of a steam yacht which had foundered on the rocks and broken up.

There is no apparent explanation for what befell the victims of this tragedy. Nor has any theory been offered as to why the lighthouse telegraph system, its one direct link with the mainland, was found to have been deliberately disabled beyond repair. But perhaps most enigmatic is the fact that a number of diamonds, accumulating to some considerable value, were found scattered down the lamp room stairs.

Of course, the truth, if it ever emerges, may well be quite mundane.

Oscar Botcherby was an expatriate Briton originally from Shepton Mallet. Why his death in Seville was of interest to UNIT remains a mystery — as does the identity of his assassin. He is pictured here with his friend Anita (surname unknown) with whom he managed *Las Cadenas* restaurant.

```
4  X  WHOLE LOBSTERS
6  X  CLAM & SQUID
4  X  BRAINS IN WHITE SAUCE
2  X  WHOLE SUCKLING PIG
1  X  WHOLE HAM WITH FIGS
8  X  SIRLOIN STEAK
1  X  FAMILY PAELLA (SERVES 12)
12 X  PIGEON BREASTS
4  X  BOTTLE RIBERA DEL DUERO
8  X  BOTTLE RIOJA GRAND RESERVA

TOTAL: PTS 81,600
```

Receipt for the meal consumed by Oscar Botcherby's murderer and his accomplice. The two men seem to have consumed a prodigious amount of food (and wine). Las Cadenas catered largely to the tourist market, hence the bill is in English. The total cost, given here in Spanish pesetas, converts to about £500.

INTERCEPT 1:

G3 intelligence. G3 intelligence. Field Major Styre reporting from Earth base.

[STATIC]

As we knew, the Earth has not been repopulated. I have therefore carried out my instructions and lured a group of humans to the planet for testing. The results of my experiment indicate that they are puny beings with little resistance to physical stress, and are totally dependent on organic chemical intake for their energy supply.

[STATIC]

Some inconsistencies have arisen. Small mistakes that will soon be eliminated. My final intelligence report for the assault will be with you within the hour...

These extracts, apparently taken from a Sontaran strategic report, appear in the UNIT files labelled 'Space Time telegraph Intercepts'.

INTERCEPT 2:

Field Major Styre, Sontaran G3 Military Assessment Survey. Experiment five, human resistance to fluid deprivation. Data, subject died after nine days, seven hours. Impairment of mental faculties, motor reflexes and physical coordination noted after only three days. Conclusion, dependence on fluid is a significant weakness which should be exploited in our attack. As a rider to the above, we should also like to take into account the successful conclusion of experiment four, where immersion in the fluid H_2O produced asphyxiation in less than three minutes. Conclusion, this species has little resistance to immersion in liquids.

The technology is Out Of This World!!

ATMOS

The ATMospheric Omission System

The revolutionary new Vehicle Management System that reduces CO2 and other emissions to ZERO

Easy to fit on any vehicle – just ask at your next service or MOT

Includes FREE SatNav and Entertainment System as standard. Improves fuel efficiency and monitors driver behaviour.

ATMOS – making our planet a cleaner, safer place to live

A product of Rattigan Industries: Driven by Design

The last notes made by reporter Johanna Nakashima before her tragic death in a freak car accident.

Deaths in vehicles fitted with ATMOS –

(52) yesterday!

5am in UK, 6am – France... Same time?

8am in Moscow. 1pm China – ALL DEATHS AT EXACTLY THE SAME TIME.

Call UNIT? Need more proof. Rattigan is hiding

something.

Plan:

- See Rattigan
- Then go to UNIT
- They have to do something.
- ATMOS is DANGEROUS.

Latest advertising blurb says ATMOS fitted to half of all vehicles IN THE WORLD. That's 400 MILLION ATMOS vehicles!

What if they are all dangerous???

LIEUTENANT

COMMANDER

GENERAL

The ranks of otherwise identical Sontarans are denoted by insignia embossed on the helmet 'collar'.

"There aren't many people as clever as me in the world…"

wonder if it's difficult for someone so young to adapt to fame and fortune. Again, we get the slight pout of the lips, as if the fact I even have to ask the question is somehow beneath contempt.

"It's not new to me," Luke Rattigan says. "I mean, I was developing professional-level software when I was 10 years old. International Electromatics offered me a job when I was 11."

So why didn't he take it, I wonder?

"I'd already got the idea for the Fountain Six search engine. It bases the results on a heuristic rather than a literal query raised against a relational database that's generated from the main internet hubs. Then an AI engine parses the query in terms a knowledge-based system can understand and returns the page hits."

"So you're more likely to get the results you want?" I say.

He looks at me like I'm about 6. "Right."

"Is it easy being so clever?" I'm joking, of course. And maybe a bit intimidated by his attitude. But he takes the question seriously.

"God, no. That's why I set up the Rattigan Academy. There aren't many people as clever as me in the world…"' I think he said "many", but it might have been "any". I nod and force a smile as he goes on: "So the Academy is a way to reward that genius and make the 0.0001 per cent of people who have the ability to join us feel appreciated. And if we can harness our skills to develop new technologies, we can advance the human race."'

Does he mean like the revolutionary ATMOS system, I ask? And immediately the scowl is back, the petulant whine of the child that, let's face it, he still is: "It stands for Atmospheric Omission System. You can't say 'ATMOS system' because that's like saying 'Atmospheric Omission System system'. It doesn't make sense, does it?"

There are times when it feels like I'm talking to a stroppy teenager who would rather be doing something – anything – else, rather than to a multi-billionaire genius who's changed the world. But I let it go and we move on to talking ab

This article by Jeremy Fitzoliver in *Metropolitan* was one of the last interviews given by Luke Rattigan before the ATMOS Incident and his subsequent disappearance.

Images taken from security cameras at the main ATMOS facility on the outskirts of London. Are these the Sontarans?

What we know about the Sontarans comes from data contained in various leaked UNIT files and documents. As always there is a danger that some or all of this information may have been tampered with or fabricated. However, the information does match what can be derived or inferred from other sources.

Coming from the high-gravity planet Sontar, the Sontarans are a clone race. They can reproduce at a rate of millions in seconds at enormous muster parades. It was to enable faster and more populous cloning that the Sontarans wanted to change Earth's atmosphere – providing sustenance for emerging clones.

The reason the Sontarans need so many new clones is because they are a race dedicated to warfare. Everything they do, even their manner of speech and the way they think, revolves around a fascination with war. Along with that comes a highly developed – if rather bizarre – sense of honour. The Sontarans have been at war with their sworn enemy the Rutans for over 50,000 years, and the war shows no sign of abating.

According to some, the Fang Rock incident was the result of an incursion by the Rutans rather than the Sontarans, who realised our planet's strategic value before the Sontarans did. Data in the UNIT files that purports to have been provided from the far future describes a time when the Sontarans too will attempt to take planet Earth for purely strategic reasons, first making a detailed assessment of their potential enemy – humans.

Methodical and thorough, the Sontarans are highly advanced technologically, with powerful warships and scout spheres that can deploy awesome weaponry. As well as osmic projection, the UNIT files describe a Cordolaine signal that the Sontarans used to make conventional ballistic weapons useless, by expanding the copper jackets of bullets so that guns in effect jammed.

The Sontarans rarely enter into alliances with others, and when they do it is invariably on their own terms. It seems likely that they were quite prepared to betray Luke Rattigan once his use to them was over. UNIT data makes mention of the Vardans – a race that can travel along any broadcast wavelength to materialise at its end point. The Vardans also apparently helped the Sontarans, only to be abandoned once they had served their strategic purpose.

If the Sontarans have a weak point, apart from their narrow-minded confidence in their own abilities, it is their Probic Vent. This is a small 'socket' at the back of the neck, through which a Sontaran ingests pure energy as sustenance. While it has been suggested that the recharge equipment could be sabotaged so that the energy instead fed on the Sontaran, there is a far simpler way of disabling the creature. The Probic Vent needs to be exposed (possibly for ventilation as well as access to energy charging), and is extremely sensitive. A blow to this area can incapacitate – and in extreme cases kill – a Sontaran.

Typically, the Sontarans claim this is not a weakness, but rather a strength – as it means they must always face their opponents in battle.

BEHIND THE SCENES

Robert Holmes wrote his outline for the story that would become *The Time Warrior* (originally titled *The Time Fugitive*) in the form of a military citation for bravery, describing the events of the story and purportedly sent from Hol Mes to Terran Cedicks (script editor Terrance Dicks). Holmes was not keen on the medieval setting that Dicks had specified, and played up the science fiction rather than historical aspects of the story.

The job of designing the single Sontaran – Commander Linx – was a collaboration between make-up designer Sandra Exelby, costume designer James Acheson and the BBC Visual Effects department, which provided the mask sculpted by John Friedlander. The design picked up on a line in the script which described Linx as having a toad-like face – a description also echoed in the dialogue as Irongron nicknames him 'Toadface'. The result was an alien with no apparent neck, and whose head was the same shape as his helmet.

The second Sontaran story, *The Sontaran Experiment*, again only required one costume as actor Kevin Lindsay, who had played

Linx, returned to play both Field Major Styre and his superior the Marshal (seen only on a video screen). The story was shot entirely on location using Outside Broadcast video equipment, and a new, more lightweight costume and mask were created. Even so, Kevin Lindsay, who was not in good health, found wearing the heavy, hot costume extremely strenuous and was doubled in the fight scenes by stuntman Stuart Fell. The Doctor was doubled extensively by another stuntman Terry Walsh after Tom Baker broke his

collar bone while shooting the story. His sling was hidden in some scenes by the Doctor's long scarf.

One inadvertent change in design was that Styre was given a full set of five fingers rather than the bifurcated two fingers and thumb that Linx had.

In their third story, *The Invasion of Time*, the Sontarans were back in force – and invading the Time Lords' home planet of Gallifrey. Four Sontaran costumes were created for the story, but just the one mask for their commander, Stor – played by Derek Deadman.

The Sontarans' final appearance in the classic series of *Doctor Who* was in *The Two Doctors*, which saw the Second Doctor captured by the Sontarans and the Sixth Doctor coming to his aid. Two Sontarans were created for the story, which was largely shot in Spain. The costume design was slightly revised, partly to take into account the temperature on location. The two Sontarans – Group Marshal Stike (played by Clinton Greyn) and his subordinate Varl (Tim Raynham) both had face masks, but the difference in their heights – and the fact that both were taller than previous Sontarans – played against the notion that all Sontarans are identical clones.

The cloned nature of the Sontarans was a major plot point when they returned for the two-part story *The Sontaran Strategem* and *The Poison Sky*. As well as creating a cloned copy of Martha, the Sontaran plan is to convert Earth into a Clone World where new Sontarans can be produced in their millions.

The story also saw the largest number of Sontarans to date as they did battle with the forces of UNIT – with ten costumes produced. Once Christopher Ryan was cast as the Sontaran commander, General Staal of the Tenth Sontaran Battle Fleet, all the other Sontaran actors were chosen for their similar height.

The Sontarans' armour was updated and changed to a blue colour. A full-sized sculpture

71cm

of the new armour was made on a body-cast of actor Christopher Ryan. From this fibreglass moulds were taken for the various components, and the suits were then made from foam latex. Each had to be adapted slightly for the varying builds of the different actors.

Two masks were produced, one for Staal and the other for his deputy Commander Skorr, played by Dan Starkey who is now better known now for playing Madame Vastra's Sontaran manservant

THE TIME WARRIOR
We see that the door of the spaceship is beginning to open.
IRONGRON and the OTHERS watch in amazement as LINX appears.
LINX wears Sontaran space armour, not unlike a gleaming,
sophisticated version of the knight's armour of the period.
IRONGRON: A warrior. A warrior from the stars. Do you come
to challenge me, sky warrior?
IRONGRON advances, sword raised.
LINX produces some kind of hand weapon from his belt and
points it at IRONGRON. There is a glow and a buzz (or
whatever) and the sword flies from IRONGRON'S HAND.
General amazement.

Strax. Both were made, as were the costumes, by the Millennium FX company.

The two Sontaran heads were formed from an overall headpiece, with the face as a separate prosthetic appliance. This was added afterwards and blended into the main headpiece. This meant that not only could the actor's eyes and mouth be seen through the prosthetic, but expressions were also transferred through it to give a very lifelike appearance.

It took about three hours for a 'full' Sontaran to get into the whole costume and make-up – far longer than previously. The Sontarans in *The Two Doctors* took about half an hour to get prepared. But the impressive final results for the new Sontarans were well worth the time spent.

SIGNIFICANT SCREEN ENCOUNTERS

THE THIRD DOCTOR

THE TIME WARRIOR
Written by Robert Holmes
Directed by Alan Bromly
First Broadcast: 15 December
1973–5 January 1974

THE FOURTH DOCTOR

THE SONTARAN EXPERIMENT
Written by Bob Bake and Dave Martin
Directed by Rodney Bennett
First Broadcast: 22 February–1 March 1975

THE INVASION OF TIME
Written by David Agnew
Directed by Gerald Blake
First Broadcast: 4 February–11 March 1978

THE SIXTH DOCTOR

THE TWO DOCTORS
Written by Robert Holmes
Directed by Peter Moffatt
First Broadcast: 16 February–2 March 1985

THE TENTH DOCTOR

THE SONTARAN STRATAGEM / THE POISON SKY
Written by Helen Raynor
Directed by Douglas Mackinnon
First Broadcast: 26 April–3 May 2008

BEHIND THE SCENES

It wasn't only the Sontarans themselves who got a makeover for *The Sontaran Stratagem*, but also their spacecraft. In previous stories, the distinctive 'golf ball' shaped Sontaran scoutships were rarely seen properly in flight. The effect of Linx's ship crashing in *The Time Warrior* was achieved with a simple lighting effect, while models were used in *The Two Doctors* as the Sontarans attacked Space Station Camera.

For *The Sontaran Stratagem* and *The Poison Sky*, both the main Sontaran Warship and their spherical Scout Pods were computer-generated images designed and created by The Mill. ●

THE SONTARAN STRATAGEM
From the shadows steps SONTARAN GENERAL STAAL; short, stocky, strong, in full uniform, including domed helmet. A strutting, formal, military General through-and-through, complete with swagger stick.

The Weeping Angels

IMAGINE IF STATUES could move. It sounds ridiculous, but down the ages mythology and legend are scattered with tales of people and animals turned to stone, and of statues coming to life. There is something fundamentally unsettling about representations of life that are themselves inert. But is this because, deep in our darkest imaginings, we fear that something so apparently lifelike could in fact be alive? And is that because of our fear of the opposite situation – of ourselves being turned to stone?

What if the stories and myths are based on truth? Is the Medusa story derived from an actual monster that could turn you to stone with a single glance? Is the legend of Don Juan – immortalised by Mozart, amongst others, in his opera Don Giovanni – picking up on an actual event? In the story, Don Juan invites a man he has murdered to dinner – and the man's statue accepts the invitation, condemning Don Juan to damnation when he arrives...

There are two theories as to how this could be possible. The first is that the stone structure of a statue can somehow be imbued with life – animated by some scientific or occult process. This is the notion advanced by the Devil Enders Society, who use reports of a grotesque stone statue of an impish demon coming to life and apparently terrorising the inhabitants of a Wiltshire village during the 1970s. It has to be said that the society's beliefs on this and other subjects are rather tenuous and implausible, ranging as they do from stories of death and destruction emanating from an ancient burial mound to alleged sightings of the devil himself.

A second, and perhaps more plausible, theory is that not all statues are actually statues at all. There is, it is suggested, a life form that is actually made of stone. What seem to us to be statues may in fact be living creatures that, for reasons that are not fully understood, cannot (or do not) move when anyone is watching.

The theory may be incomplete in scientific terms, but it does go some way to explaining reports of statues that apparently change their position, or even their location. Could it be that the story of Don Juan and countless other 'living statue' stories, from Ovid to Shakespeare, Schalcken to Wilde, are all derived from some innate knowledge of this strange and potentially dangerous life form?

The unexplained disappearance of Billy Shipton was an embarrassment to the police. Not only was he a detective inspector, but at the time that he mysteriously vanished, he was investigating a number of missing persons cases. Shipton believed the cases were all connected.

His investigations had centred on an abandoned, derelict house called Wester Drumlins. Once a favoured location for couples to meet clandestinely, the house had gained something of a reputation for being haunted. Witnesses reported seeing strange figures in the grounds, or at the windows of the house. According to some, these figures were angels. But several witnesses who got closer to them described one or more stone statues – angels weeping into their hands.

But D.I. Shipton wasn't so interested in the people who returned from Wester Drumlins with stories of spooky statues or mysterious figures. His concern was for the people who never came back. With the growing reputation of the 'Scooby Doo House', as it was nicknamed locally, a lay-by on the road outside became the new haunt (if that is the right word) for young couples. Shipton's suspicions were raised, and his enquiry launched, when three abandoned cars were found in this lay-by in less than a month. These were not vehicles that had been dumped after failing their MOT test. There was no explanation for why they were left, and the occupants were never found. In all, Shipton's team collected 17 vehicles from the area – and also, according to some reports – an antiquated police telephone box.

Wester Drumlins — a haunted house?

DON'T BLINK

Is this man photographed in 1969 (left) actually Billy Shipton?

Reclusive 'businessman' Julius Grayle disappeared from his house in Manhattan

Sam Garner — did he investigate the wrong mystery?

DON'T BLINK

Shipton himself seems to have been the last victim to disappear, following close behind Miss Katherine Costello Nightingale. What happened to these people? Where did they go? Were the weeping angel statues at Wester Drumlins somehow responsible?

The answers, bizarrely, may lie in 1930s New York, where a spate of similar disappearances were also linked – at least in local rumours and gossip – to statues of weeping angels. The disappearances are of people as diverse as private detective Sam Garner and the notorious, reclusive 'businessman' Julius Grayle. Again, the stories and the disappearances all seemed to centre on a single building, in this case the Winter Quay apartment block. Just like Wester Drumlins, this was a location avoided by locals and said to be haunted. The building no longer exists, but some photographs from the period clearly show a pair of statues at the entranceway – statues of angels. In other photographs, worryingly, the statues are not present...

Reports that the Statue of Liberty was somehow also implicated remain unsubstantiated and implausible. But is the Curse of the Weeping Angels merely an urban legend, that has somehow migrated from 1930s Manhattan to modern Britain? Or is it something altogether more dangerous?

MELODY MALONE

Some attribute the origin of the Weeping Angels stories to the 'Melody Malone' series of pulp crime novels. From this extract from 'Heart of Stone', it's easy to see why. There is some evidence to suggest that Melody Malone may have been based on a real character.

PRIVATE DETECTIVE IN OLD NEW YORK TOWN

...didn't tell me. Maybe he was afraid to say. There were a lot of things he was afraid to say. He was even afraid to look me in the face, though that didn't stop him looking at me everywhere else. I greeted him like the old friend that he wasn't and led him through to the Long Gallery.

'What's with the pictures?' Lassiter finally asked.

'Well, there are frames with most of them,' I said. That seemed to confuse him and I was treated to another stretch of silence. I didn't let it last for long, because I knew he liked the sound of my voice. Also, I like the sound of my voice. In fact, I don't think I know anyone who doesn't. At least, no one worth speaking to.

'We're not here to look at pictures,' I told him,

TAKING A LIBERTY!

While the island of Manhattan is awash with stories of missing persons, Jeff Loadster of Pentallion Drive reported the most spectacular case of a missing person so far – the Statue of Liberty.

"Hey," he told our reporter, "I looked out my window at about 3am and there it was – gone." Loadster has a good view out towards Liberty Island from his back window. "It's a heck of a sight, the statue against the sky across the river. And that night, it just wasn't there. Blowed if I know where the lady went. It ain't like I'd been drinking or anything. Not much, any road."

We understand that several other people reported the statue "missing" but NYPD refused to comment except to say there is no ongoing investigation into the incident.

One of the more bizarre stories published by the New Jersey Enquirer in 1930

---

adjusting my décolletage. It probably wasn't much o a disappointment, as he wasn't looking at the pictures anyway. But enough about me.

'So, what's the deal then?'

'Statues,' I said.

'Statues,' he echoed, for all the world like a huge, slightly balding, slightly sweaty parrot dressed in a shabby suit and down-at-heel raincoat.

'Statues,' I agreed. 'One statue in particular. She's just through here.'

'A dame?' Lassiter perked up at that. If 'perk' is the right word.

'No one knows who sculpted it,' I explained as we reached the statue in question, and it was a big question. 'Though my guess is it was never sculpted at all. But it's called "She Weeps".'

The stone angel hid her face behind her hands. But I was pretty certain she knew we were there. I was pleased to see the exhibit was roped off. Even so, Lassiter couldn't resist reaching out to touch the cold, brittle stone. I slapped his hand away, and pointed to a sign on the plinth. 'It says don't touch.'

'That's just for, like, insurance or something,' he said. He really had no idea.

'That's just for saving your life or something,' I told him. 'Touch the statue, and you're history. And I mean that, Sweetie.'

WINTER QUAY

Last few Luxury Apartments available
Lifetime Guarantee on all Fixtures & Fittings
Specially suitable for young people with no dependents
Reasonable Rent and Low Maintenance
WINTER QUAY – A PLACE TO SPEND THE REST OF YOUR LIFE

DON'T BLINK

Receiving no answer, we proceeded inside. There was evidence of a struggle: broken windows, tables overturned, broken ornaments etc. I organised teams to search the residence, which is a substantial town house, but there was no sign of Julius Grayle.

I don't need to state how disappointed and frustrated everyone felt. Just as it seemed we finally had the evidence to bring Grayle to book for his crimes and misdemeanours, he had somehow got wind of our operation and fled.

Chief O'Hara's report details the extent of the state-wide search that we instigated, and the subsequent federal investigations. But to this date there have been no credible sightings of Julius Grayle and the case file remains open.

There is one other rather odd thing to report from that evening. Questioning neighbors and other local residents to try to ascertain Grayle's movements, several pointed out that a statue of a mother and child that was prominently situated in front of the house has disappeared. I have filed a report on behalf of the residents, though one woman did remark: "I never liked the thing. Creeped us all out, I tell you. You'd swear the figures moved when you weren't looking."

People don't understand time. It's not what you think it is...

Complicated...

Very complicated...

People assume that time is a strict progression of cause to effect, but actually from a non-linear, non-subjective viewpoint, it's more like a big ball of wibbly wobbly, timey-wimey stuff...

It got away from me, yeah...

Well, I can hear you...

Well, not hear you, exactly, but I know everything you're going to say...

Look to your left...

I've got a copy of the finished transcript. It's on my autocue...

I told you. I'm a time traveller. I got it in the future...

Yeah. Wibbly wobbly, timey-wimey...

What matters is, we can communicate. We have got big problems now. They have taken the blue box, haven't they? The angels have the phone box...

Creatures from another world...

Part of a transcript of the speech given by the 'Easter Egg Man' who appears hidden on 17 apparently random DVDs. No one understands what the man is talking about, but the mention of 'angels' suggests a link to the statues. The implications of the full transcript are discussed later...

Letter attributed to Kathy Nightingale — who went missing from the vicinity of Wester Drumlins in 2007. Much of the detail in the letter has been corroborated, but of course the implications are incredible.

Are these all just statues?

My dearest Sally Sparrow.

15th March, 1987

If my grandson has done as he promises he will, then as you read these words it has been mere minutes since we last spoke. For you.

For me, it has been over sixty years. The third of the photographs is of my children. The youngest is Sally. I named her after you, of course.

I suppose, unless I live to a really exceptional old age, I will be long gone as you read this. Don't feel sorry for me. I have led a good and full life. I've loved a good man and been well loved in return. You would have liked Ben. He was the very first person I met in 1920.

To take one breath in 2007 and the next in 1920 is a strange way to start a new life, but a new life is exactly what I've always wanted.

My mum and dad are gone by your time, so really there's only Lawrence to tell. He works at the DVD store on Queen Street. I don't know what you're going to say to him, but I know you'll think of something. Just tell him I love him.

Kathy

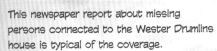
This newspaper report about missing persons connected to the Wester Drumlins house is typical of the coverage.

SPIRITED AWAY

Notorious local 'Ghost House' Wester Drumlins last week claimed yet more victims, as Andy Kenyon and Carol Martindale (both 19) failed to return from a night out. Kenyon's car was found abandoned in a lay-by close to the house – a stopping point referred to locally as 'the AWOL stop' as so many people have apparently gone missing there.

Although no direct connection with Wester Drumlins has been found, the house has a reputation for strange events. It has been empty since the last occupants, Mr and Mrs Satchwell, disappeared in the late 1960s. The house and grounds, now derelict and overgrown, have a reputation for being haunted. This notoriety has been enhanced not just by the number of people who have gone missing in the area.

The police and the local authorities have tried various methods of dissuading trespassers. The local council has even erected a warning notice declaring the house to be an 'unsafe structure'.

Despite this, local youths seem to take delight in moving the property's various ornamental statues around its grounds. One dilapidated statue of an angel figure has been reported in several different locations, for example.

Detective Inspector William Shipton, leading the missing persons investigation, again appealed for information, as well as urging people to stay clear of Wester Drumlins. "We don't know exactly what has happened to Andy and Carol," he said in a statement on Wednesday. "But until we are sure they are safe and sound, we have to assume the worst. I'd urge the public not to trespass on the land at Wester Drumlins, and to keep away from the area unless accompanied. We cannot rule out the possibility of foul play."

DON'T BLINK

DON'T BLINK DON'T BLINK

The most intriguing and perhaps enlightening clues about these 'living' angel statues comes from the most unlikely of sources. News of the 'Easter Egg Man' first surfaced on internet forums and conspiracy websites several years ago. Video of this unidentified bespectacled man was found hidden in 'Easter Eggs' – extra material available from concealed menu choices on a DVD – speaking to the camera.

Who this man (or the young woman briefly glimpsed with him) might be, or how he came to be on a total of 17 apparently random DVDs has never been explained. The DVD manufacturers claim to have no knowledge of how the video – completely unconnected to the other contents of the DVDs – came to be included. But it is what the man says that is especially fascinating.

Much of it seems to be random gibberish, including a rambling discourse on the nature of Time. But one section has particular resonance:

- The lonely assassins, they used to be called. No one quite knows where they came from, but they're as old as the universe, or very nearly, and they have survived this long because they have the most perfect defence system ever evolved. They are quantum-locked. They don't exist when they're being observed.

- The moment they are seen by any other living creature, they freeze into rock. No choice. It's a fact of their biology. In the sight of any living thing, they literally turn to stone. And you can't kill a stone. Of course, a stone can't kill you either. But then you turn your head away, then you blink, and oh yes it can.

- That's why they cover their eyes. They're not weeping. They can't risk looking at each other. Their greatest asset is their greatest curse. They can never be seen. The loneliest creatures in the universe.

Is this an explanation for the countless tales of living statues? Does it help to explain the people missing from the area of Wester Drumlins? The classic test of a viable theory that would satisfy the critics is whether it can explain the Stockinsky Painting.

Ivan Stockinsky was a Russian painter who fled to London after the Russian Revolution, and became an outspoken critic of Soviet Communism. Best known, of course, for his Study of Light and Death, a lesser known work is entitled Glimpsed in a Graveyard. This 1938 painting takes as its subject the tomb of Edwin Manners in Highfield Cemetery. Stockinsky was noted for his attention to detail and the accuracy of his paintings, especially the landscape. Yet in this painting, one of the 'Weeping Angels' that stands vigil outside the tomb appears to have taken on a more menacing aspect.

DON'T B

Ivan Stockinsky's notorious painting 'Glimpsed in a Graveyard'. Does it hold a clue as to what happened to the artist?

Did Stockinsky, in this one painting, depart from the style and habit of a lifetime? And if so, why?

The only man who would know for sure was Ivan Stockinsky himself. But sadly, this was his last work. The painting was found, recently completed and still on an easel set up in front of the tomb that forms its subject. There was no sign of Stockinsky, and despite a thorough search by the police, he was never found. The accepted theory is that he was kidnapped and then assassinated by Russian agents under the orders of Stalin.

But does this painting suggest a different explanation? Intriguingly, if you visit the tomb today, there is only one angel outside. What happened to the other has never been explained.

DON'T BLINK   DON'T BLINK

There is a classic conundrum: if a tree falls in the forest and there's no one there, does it make a sound? Whether you call them the Lonely Assassins or the Weeping Angels, they pose a similar riddle: If a statue moves when no one is looking, does it really move? Certainly there are instances where statues have changed position or even location. But whether they did it without the intervention of others is impossible to say. The myths and stories of living statues are just that – myths and stories. For all his apparent authority and knowledge, the Easter Egg Man could be just an actor speaking lines written by someone with a vivid imagination…

But there is no denying that those lines, those words of warning, speak to something deep in our collective psyche. There is an old children's game called Grandmother's Footsteps. It involves one person turning away while others try to sneak up on them. But when they turn round, everyone else freezes – if they see anyone move, that person is out of the game. Could this harmless playground entertainment actually have its origins in truth? Was it first played as a way of educating our children about the dangers of the Weeping Angels?

If what the Easter Egg Man tells us is true, then there is only one defence against these creatures. In the words of the mysterious man himself:

They're coming. The angels are coming for you. But listen. Your life could depend on this. Don't blink. Don't even blink. Blink and you're dead. They are fast. Faster than you can believe. Don't turn your back. Don't look away. And don't blink. Good luck.

Could this man's hidden words save your life?

BEHIND THE SCENES

The Weeping Angels first appeared in the episode *Blink*, written by Steven Moffat. Having already contributed scripts to the first two series of the revived *Doctor Who*, this time Moffat faced an additional challenge. *Blink* was to be made in parallel with another episode, as a way of saving time in the production schedule. That meant that actors David Tennant and Freema Agyeman would only be available for a few short 'insert' scenes as the Doctor and Martha, interspersed with the main action of the story. The same technique had been used the previous year for the episode *Love & Monsters*, in which the Doctor and Rose also only appear briefly.

The story for the episode came primarily from two sources. One was a short story called *What I Did On My Christmas Holidays by Sally Sparrow* – which Steven Moffat had written for the 2006 *Doctor Who Annual*. The story told how 12-year-old Sally found a message from the Doctor hidden beneath the wallpaper asking for help. She also found a message from the Doctor on a videotape – in which he told her he was stranded in 1985 after the TARDIS went wrong and left him there. The Doctor has a copy of her essay from the future which is how he can keep up his side of the conversation on the tape. Of course, Sally takes the TARDIS back to rescue the stranded Doctor.

The short story didn't include any villains, and inspiration for the Weeping Angels came from another source. Partly, they were derived from the playground game Grandmother's Footsteps, which Moffat confesses he always found a bit frightening. In this, players have to

THE TENTH DOCTOR

BLINK
Written by Steven Moffat
Directed by Hettie MacDonald
First Broadcast: 9 June 2007

THE ELEVENTH DOCTOR

THE TIME OF ANGELS / FLESH AND STONE
Written by Steven Moffat
Directed by Adam Smith
First Broadcast: 24 April–1 May 2010

THE ANGELS TAKE MANHATTAN
Written by Steven Moffat
Directed by Nick Hurran
First Broadcast: 29 September 2012

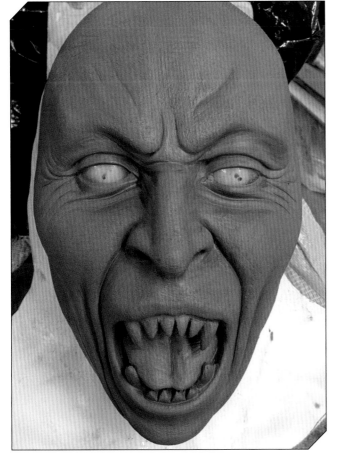

BEHIND THE SCENES

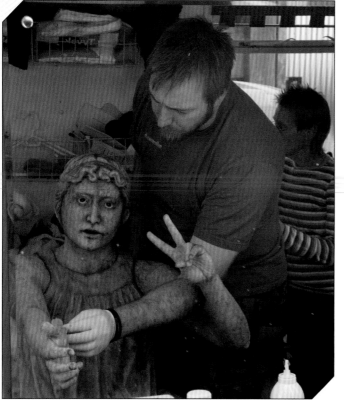

creep up on someone from behind – and freeze to immobility when the person turns to look. Anyone who is seen moving is out of the game. The winner is whoever can creep close enough to touch the person they are creeping up on…

In the version of the game played in *Blink*, the Weeping Angels always win – and the consequences are truly frightening. So frightening that the Weeping Angels returned in the Eleventh Doctor's first series in the two-part story *The Time of Angels / Flesh and Stone*. In this, their nature was refined and developed – with an Angel even escaping from a security video to menace Amy.

Amy, together with her husband Rory, met the Weeping Angels again in their final story, *The Angels Take Manhattan*. In this they were menaced by the biggest Lonely Assassin of all – the Statue of Liberty.

In many ways, animating the menacing Statue of Liberty was a throwback to the first thoughts about how to create the Weeping Angels, as it was achieved entirely by use of computer-generated imagery (CGI). This had been proposed as a way of creating all the Angels for *Blink*, but would have been costly and difficult given the number of sequences needed.

BLINK
AN ANCIENT STONE STATUE, standing tall and thin and solemn in the overgrown garden —
ancient, weather-beaten, stained and mottled by a hundred years of rain.
Its head is bowed, and its face is buried in its hands, like it's lamenting.
Or weeping…
Weeping! Sally looks back to the first line of the wall writing —
BEWARE THE WEEPING ANGEL

Producing actual statues was also considered, but given the number of different poses needed, that too was deemed too complicated and expensive. Instead, the production team returned to a solution they had initially discarded as too tricky to implement – using actors in make-up and costume standing very still. To overcome the main perceived problem and make sure that the Angels didn't 'wobble', the actresses had tall stools, like shooting sticks, hidden beneath their robes on which they leaned to maintain their statuesque static position.

The angels' skirts were made from rigid polyfoam, sculpted by the team at Millennium FX, who also provided the lightweight, flexible silicon mask, as well as make-up for arms, polystyrene wings, and 'shells' to cover the angels' eyes.

The robes were created from real fabric. This was positioned on a mannequin then painted with polyester resin to make it rigid, and enhanced with detailing made from clay. A mould could then be taken, from which a foam latex bodysuit was made. The wings were extremely lightweight and attached to a special corset worn by the actress, which helped ensure they stayed rigidly in place.

Any bare skin – for example on the arms and neck – was treated with make-up to look like stone and blend in with the costume. Several layers of make-up were needed, each dried with a hairdryer before the next was applied. In all, it took about three hours for each performer to be transformed into an angel. ●

THE TIME OF ANGELS
THE DOCTOR
You're still here.
(Rounds on her)
Which part of "wait in the TARDIS till I tell you it's safe" was so confusing?

AMY
Ooh, are you all Mr Grumpy-Face today?

THE DOCTOR
A weeping angel, Amy, is the deadliest, most powerful, most malevolent life form evolution has ever produced – and right now one of them is trapped inside that wreckage and I'm supposed to climb in after it, with a screwdriver and a torch, and assuming I survive the radiation long enough, and assuming the whole ship doesn't blow up in my face, do something incredibly clever which I haven't actually thought of yet. That's my day, that's what I'm up to, any questions??

AMY
Is River Song your wife?

The Zygons

LOCH NESS, NEAR INVERNESS in the Highlands of Scotland, is the greatest mass of fresh water in the British Isles and forms part of the Great Glen. The loch is 23 miles long, nearly 800 feet deep, and covers over 700 square miles. Its size helps to explain how it can be so difficult to prove or disprove the existence of the Loch Ness Monster.

The first recorded sighting of the monster was in AD 565 when St Columba visited Loch Ness. The historian Adamnan (later Abbot of Iona) wrote: 'Of the driving away of a certain water monster by virtue of prayer of the holy man' and tells how the saint sent the creature back to the depths of the loch.

Though there were reports of a creature in the loch in the late nineteenth century (notably in 1880), it was not until 1933 that sightings really began to pick up. Until then, the loch was largely obscured from sight by trees and vegetation. But in 1933 a track on the north of Loch Ness was made up into a road. This involved considerable blasting with dynamite as well as the uprooting of vegetation at the water's edge. This had two possible effects. First, it generated a lot of noise and deposited debris into the water, the combination of which may have disturbed whatever lives in the loch. The new road also provided good views of the water. By mid-1933, monster-sightings were becoming common.

In the late 1970s, the Loch Ness Monster was blamed for the mysterious destruction of several oil rigs – despite the fact that they were situated not in the loch but the North Sea. Even further afield, 'Nessie' was linked to reported sightings of a 'long-necked dinosaur-like creature' in the River Thames.

But are these claims as bizarre as they at first appear? Or could it be, as some sources claim and secret UNIT files seem to imply, that the Loch Ness Monster is actually the 'pet' of a colony of aliens who have lived undisturbed in Loch Ness for centuries?

According to UNIT, a Zygon spacecraft crash landed in Loch Ness centuries ago. The stranded Zygons brought with them an embryonic creature – the files call it a Skarasen – that grew to provide the Zygons with sustenance as well as acting as a defensive weapon as they remained hidden for centuries in the loch. But then something changed, and the Zygons went on the offensive…

Answers to the mystery of Loch Ness may lie not in the loch itself, but in the nearby village of Tulloch (sometimes spelled 'Tullock'). At the heart of Tulloch Moor, the village comprises a few houses, a village shop, a local inn called the Fox, and several outlying farms. Yet despite its size Tulloch has been at the centre of several unexplained events over the past few hundred years.

Tulloch Moor is notorious for its thick fog, which can descend over the landscape in minutes. It is a fog in which people have vanished, and monsters have appeared. Few locals will go out on the moor after dark, and they once believed the fog was the steam from a witch's cauldron. This may also explain the name of a small lake on the edge of the moor – the Devil's Punchbowl.

Whether attributed to witchcraft or devilry, Tulloch Moor has a reputation for the unusual. The area is rife with ghost stories as well as the inevitable sightings of Nessie. But there are also stories of other creatures – the size and rough shape of a human. Are these the Zygons mentioned in the UNIT files?

It is certainly no secret that UNIT forces were in the area, based in Tulloch itself, during and immediately after the North Sea oil rig disasters. Since the oil company involved, Hibernian Oil, had its own base nearby, this is not in itself surprising. But is it coincidence that the landlord of the Fox died of a suspected heart attack while UNIT personnel were actually staying at his inn?

What mysteries lurk on Tulloch Moor?

The Duke of Forgill (left) and another man

THE ZYGONS

UNIT personnel in Tulloch

The Loch Ness monster — an alien?

Another connection, perhaps too neat to be a coincidence, involves the Duke of Forgill, whose estate borders Loch Ness on one side and Tulloch Moor on the other. In addition to his local duties, including his roles as Chieftain of the Antlers Association and Trustee of the Golden Haggis Lucky Dip, the Duke was President of the Scottish Energy Commission. In that capacity he attended the International Energy Conference in London – at Stanbridge House. It was here that the sightings of the 'London Nessie' were reported, on the day the conference opened. Some of the delegates who met the Duke that day reported that he was 'not himself'. Did Forgill know something his fellow delegates did not? Did the Loch Ness Monster, for some reason, follow its local laird down to London?

And what happened to the Zygons themselves? It is known that UNIT obtained permission to test a new type of depth charge in the waters of Loch Ness on the very same day. Some local observers reported hearing the explosions, as well as the sound of a huge aircraft, although no flights were scheduled in the area.

As usual, the information gleaned from the secret UNIT files raises more questions than it answers...

275

RIG:
Hey, listen, Willie. With tomorrow's supply ship, can you no send over a few haggis? The chef we have here doesnae ken the first thing about…

[STATIC / INTERFERENCE]

Willie? Hello, Willie, can you hear me? Over. This is Charlie Rig to Hibernian Control. Are you receiving? I say again, Charlie Rig to Hibernian Control. Do you read me? Over. Mayday! Mayday! Mayday! Mayday!

[SIGNAL OVERWHELMED BY STATIC]

HIBERNIAN CONTROL:
How are things out there? Morale okay? Over.

RIG:
Everything's fine. Any news of your investigations? Over.

HIBERNIAN CONTROL:
Nothing much. The Brigadier

[STATIC / INTERFERENCE]

RIG:
Hello? Hello? Hibernian Control, this is number three rig. Are you receiving me? I say again, are you receiving me? Over.

HIBERNIAN CONTROL:
Ben Nevis, Ben Nevis, are you there? Are you there? Over!

[SIGNAL OVERWHELMED BY STATIC]

never returned to the Black Isle. But the legends of Tulloch Moor date back to well before this 1922 tragedy.

One of the better-documented incidents occurred back in 1870. Two brothers, Robert and Donald Jamieson, were out on Tulloch Moor cutting peat when the mist came down suddenly. Both the young men knew the moor well, having been born and brought up in Tulloch village, so there was little concern for their safety. The mist can depart as quickly as it arrives, so the locals knew simply to stay where they were and wait for the air to clear.

Perhaps the Jamieson brothers were already on their way back home when the mist fell. Perhaps they thought they could find their way to the village through the gathering gloom. But whatever happened, when the fog eventually lifted, they did not return.

When it was apparent that there was a problem, a search was organised. The area where they had been peat-digging was well known, and the search party soon found the brothers' tools and the trench they had dug. 'Bricks' of peat were piled up neatly, ready to be loaded into a cart. But of the brothers themselves there was no sign.

For two days, the local people scoured the moor for the missing brothers. They finally found Robert, the older of the two, several miles from the digging. By all accounts he was in a state of extreme shock, wandering aimlessly across the moor. Despite intense questioning, he would say nothing of what had happened – or anything else. Alexander Lamont recorded in his diary that Robert Jamieson 'had about him the aspect of one who has survived the most terrible of experiences. He was pale as a freshly laundered bed sheet, and his eyes seemed drained of all colour, staring unfixedly at a point on some distant imagined horizon…'

Robert Jamieson was committed to an institution in Inverness, where he died in 1893 of pneumonia. For the rest of his life, he never spoke again.

While it is one of the best-known relating to legends of the evil spirits that supposedly inhabit Tulloch Moor, the case of the

Jamieson brothers is far from
unique. Just three years prior t
the brothers' d

The real story behind the
London Loch Ness Monster,
or an imaginative cover-up?

MONSTER HOAX

Tourists and Londoners alike were treated to a rare sight yesterday, at least according to some reports. Eyewitnesses claim to have seen a 'gigantic monster' rear up out of the Thames close to Stanbridge House – where the International Energy Conference is being held this week.

Organisers of the conference claimed that the 'monster' was a publicity stunt, organised by environmental protesters objecting to the impact of North Sea Oil. In particular they wanted to draw attention to the damage being done to the countryside in the vicinity of Loch Ness – and so they enlisted the help of their own inflatable 'Nessie'.

The Duke of Forgill, President of the Scottish Energy Commission, whose estate borders the loch said: 'It's an imaginative way to draw attention to their cause. But I have to say that, for all I complain about the oil companies' rough-necks trespassing on my own land, North Sea Oil is very much the future for the Scottish economy. My family has served the country for seven centuries, and in that time we've seen a lot of change. The secret is to adapt to the circumstances. For what it's worth,' he added, 'I have never believed in the Loch Ness Monster. It's as much a myth as space aliens, but there are no limits to human credulity.'

wary of leaving Her Majestie too long in the company of this strange physician. I therefore took it upon myself to lead a partye into the woodes in search of them. Wherein we came across the Physician in company with two others, one an elderly gentleman as I perceived and the other a mere stripling barely into his adulthood. Her Majestie was also present, but affected a demeanour most strange.

I dare not commit to this journal the details of what ensued. But suffice to saye that there was devilrie and witchcraft abroad that day. I have heard the disembodied voice of Hecate herself, yea, and seen wonders of which I would not speak. Two Queens – alike as peas in a pod. And yet, in character so different. How could we tell them aparte? And what of the ghastly creature we did spye, all covered in suckerous growths and of an orange hue, its head a domed enlargement above deep, small, malevolent eyes?

On Her Majestie's instructions we did convey the Physician and his fellows to the Tower, wherein they were incarcerated. Yet strange to relate, Her Majesty bade us leave the door to their cell unlockd. What scheme she had in mind, I cannot saye. Her Majestie then bade us leave her, and departed unto the cellarage, wherefrom the oddest soundes did ensue.

Yet strangest of all these things I do here relate was yet to come. Her Majestie did summone us to the Tower's inner courtyard wherein she had set a strange blue tent, and without this she did, upon my life, enlist the services of a reverend father. And she was marryed to her Physician. Straight way did he and all his entourage, now enlarged by addition of a comely wench of tender yeares, depart into the blue tent. With scraping and rasping as if the heavens themselves would break asunder, the tent itself did then depart. No, not aboard a cart or other bark, but into the very air itself.

Of Her Majestie's Physician we sawe no more. And she was much perturbed and did entreat us never to speake again of that day unless the worlde itself be rendered unto dust.

An extract from Lord Bentham's journal. Historians have speculated that it was intended as a fiction rather than an account of actual events. The 'Physican' and his colleagues have never been identified.

One of the paintings stored in the Under Gallery, this portrait shows Queen Elizabeth and another man. The man's identity has provoked some discussion, given the apparent intimacy between the two.

Photographs from leaked UNIT files. Are these the Zygons?

The Royal Charter, signed by Queen Elizabeth, establishing the mysterious Under Gallery (see p280)

By order of hrh Queen Elizabeth,

To those unknowable inhabitants of the England yet to come, I place my trust in your future loyalty, that you will see fit to execute my decree. By order of Queen Elizabeth the Doctor is here appointed official curator of the Under Gallery, to be summoned in the event of any crisis concerning it. And when such time arises that the crisis is averted, I ask that the Doctor be banished from this glorious isle once more and that the Under Gallery be again sealed away for perpetuity.

Elizabeth by the Grace of God of England, France and Ireland Queen and defender of the faith

Elizabeth R

Strange though it may seem, sightings of the Zygon creatures have not been confined to Loch Ness, or even to Scotland. While it seems possible, even likely, that the Zygons have existed in and around Loch Ness and Tulloch Moor for centuries, was there another colony of Zygons further south? There are accounts of remarkably similar creatures in London as far back as the sixteenth century.

Lord Bentham describes encountering one such creature during a hunting trip to the woods with Queen Elizabeth. His account is frustratingly vague as he gets sidetracked into what seems to be a philosophical discussion about how one can be sure that the queen is actually the queen. Historians tend to be dismissive of Bentham's writings, pointing to the fact that he also gives an account of the wedding of Queen Elizabeth at the Tower of London – an event which of course never took place.

Are these actually the same painting? And if so, where did the grotesque figure go? Is it possible to step out of a painting?

More convincing are the writings of Sir Henry Cadogan, who was curator of the National Gallery for a while in the mid-19th century. He describes a secret collection of paintings apparently collected by Queen Elizabeth and stored in a concealed 'Under Gallery' below the main building. He lists some of the paintings, including Gallifrey Falls and a very similar painting (or perhaps the same painting with a different title) No More.

What is especially intriguing is that Cadogan describes this and some of the other paintings as being 'so lifelike, so detailed, and with such a depth of perspective that one could imagine stepping into the picture and finding one's self in another place entirely.' His comments on the way the perspective apparently changes when the paintings are viewed from different angles has led some scholars to suggest that there was some form of three-dimensional effect applied to the paintings.

But the comment that one could imagine 'stepping into' the paintings has prompted a more frightening theory. What if these pictures are not representations of other places, but somehow windows onto the places themselves? Cadogan describes one picture showing 'an arid, dusty landscape beneath a blue sky scattered with clouds and streaked with sunlight attempting to shine through. Rocky peaks rise on either side and in the centre of the picture a strange figure strides purposefully towards the painter. The figure has two arms and two legs, but cannot otherwise be said to be human. Its ribbed body is adorned with nodules and suckers and a domed head rises from a neck-less chest. Its colour almost blends with the landscape itself. The deep-set eyes are full of menace…'

Could he be describing a Zygon? More worrying is that an inventory of the pictures held in the Under Gallery that was recently leaked to the press includes a description of this very picture – but without mention of the otherworldly figure striding towards us.

The current Curator of the secret Under Gallery.

Is it possible that the figure kept walking, and stepped out of the world of the painting and into our own? Is this where the Zygons originated? And if so, what other horrors are waiting hidden in alien artwork concealed from the world in this secret Under Gallery and other similar collections?

BEHIND THE SCENES

The focus of *Terror of the Zygons* was always intended to be the Loch Ness Monster – in fact an early working title for the story was simply *The Loch Ness Monster*. But after discussions with the BBC's Visual Effects Department, producer Philip Hinchcliffe and director Douglas Camfield decided it would be safer to switch attention to the creature's Zygons masters.

This proved to be a wise decision: when the final stop-motion animation sequences of the monster were completed, Camfield was less than impressed and used as little of it as he could. As well as the stop-motion animated model, a 'puppet' version of the Skarasen was also created, and used mainly for the shots in Episode 4 where the creature rears up out of the Thames. Other effects– including the destruction of the model oil rigs, and the Zygon spaceship – were far more effective.

The Zygons themselves were altogether more impressive than their 'pet' Skarasen. Costume designer James Acheson came up with their design after discussions with Camfield and from the script's mention of their dependence on the Skarasen's lactic fluid. 'I remember I was looking at a lot of half-formed embryos in amniotic sacs,' he recalled in 1988. Acheson created a model, and sculptor John Friedlander created

THE FOURTH DOCTOR

TERROR OF THE ZYGONS

Written by Robert Banks Stewart
Directed by Douglas Camfield
First Broadcast: 30 August–20
September 1975

THE ELEVENTH DOCTOR (WITH THE TENTH DOCTOR AND THE WAR DOCTOR)

THE DAY OF THE DOCTOR

Written by Steven Moffat
Directed by Nick Hurran
First Broadcast: 23 November 2013

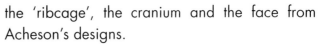

the 'ribcage', the cranium and the face from Acheson's designs.

The costumes were also fitted with lights inside the chest and head, but these didn't show up on camera and were not used. Instead, the studio lighting helped enhance the effect of the Zygons living in an underwater base.

BEHIND THE SCENES

Millennium FX, the company that has handled most of the creature construction and prosthetics for *Doctor Who* since it returned in 2005, was responsible for the design and construction of the new Zygons for *The Day of the Doctor*. Their brief was to stay close to the original, distinctive designs, but to make the updated creatures 'meatier' – more muscular and powerful, and if possible more scary.

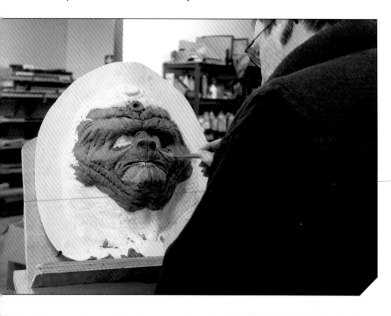

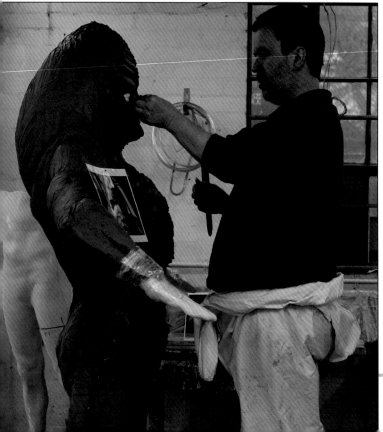

The initial concept design for the new Zygons was done in 3-D computer modelling software package ZBrush by Dave Boneywell and was close to the original costume, but updated to make the Zygons appear more muscular. The main difference from the original 1970s Zygons was at the back. Instead of a protruding 'ribcage' which gave the original Zygons their slightly hunched-over appearance, the new Zygons had extra suckers and pronounced shoulder blades and muscles.

Next, a full-sized sculpture of the Zygon was created in clay over a full-sized human mannequin. Moulds were then taken from the sculpted figure for the various components that would form the finished Zygon suit. The exception was the face – left blank on the sculpted figure – which was to be a separate make-up prosthetic.

The main part of the costume was the torso, which included the full head and both arms to the elbow. Each forearm and each leg was then another element. The hands were included as part of the forearm, which overlapped the elbow so the join could be hidden, while the feet were separate again. Each of the components was made from foam latex. The final suit was hot to wear – especially the bulky head. The top of the head was formed from a skin of very thin foam latex over a foam core with holes drilled into it to provide some cooling for the actor. The actors could also remove the hands and feet fairly easily, though the rest of the costume had to stay on all day. With the enlarged head, the complete suit ended up being approximately 6'4" tall.

Two identical Zygon suits were made, although early plans were for just one. Being created from the same moulds, both were identical. The only real differences were visible in the two sets of sharp, false teeth provided by the Fangs FX company.

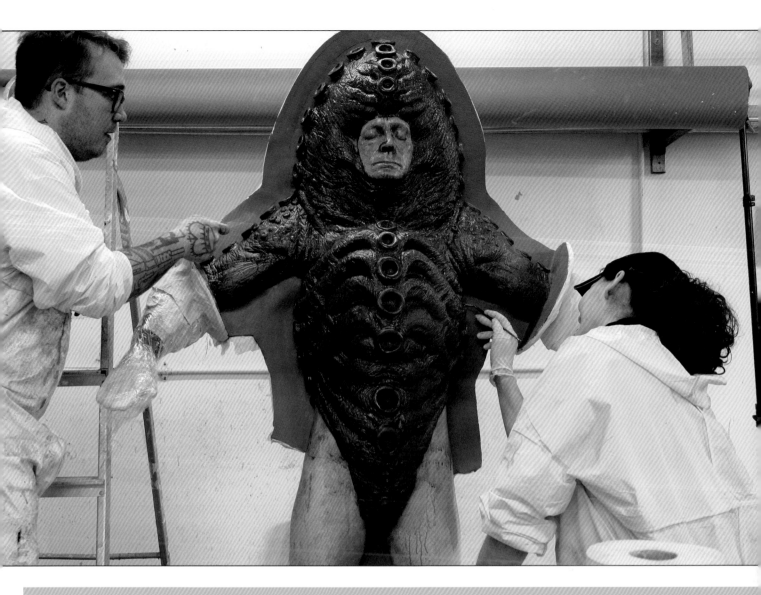

ELIZABETH
My love, I do not understand.

THE TENTH DOCTOR
I'm not your love, and yes you do! You're a Zygon.

ELIZABETH
A Zygon??

THE TENTH DOCTOR
Oh, stop it, it's over. A Zygon, yes. A big red rubbery thing, covered in suckers,
surprisingly good kisser. Do you think the real Queen of England would just decide
to share the throne with any old handsome bloke in a tight suit, just cos he's got
amazing hair and a nice horse??

As he says this he gestures to the horse -

- and double-takes.

Because standing where the horse was, is a man-shaped rubbery, red thing, covered
in suckers - a Zygon!

BEHIND THE SCENES

The final component was the Zygons' faces. These were sculpted in Plastilene by Gary Pollard on a fibreglass life cast taken of each of the two performers. The prosthetic ran from forehead down, and across from cheekbone to cheekbone. This prosthetic appliance was the only part of the suit that was not reusable, and so a new one was needed for each day of shooting.

As foam latex is an opaque material, the designers tried to give a translucent quality, rather like a squid, in the paint finish. A concept colour scheme was developed at Millennium FX using Adobe Photoshop. Henrik Svenson then did a full-scale paint test using layers and washes of paint to give a sense of depth. It took about a week to paint each suit.

Once in the suits, the prosthetic face and the various foam latex components had to be blended seamlessly together. For the actors to get into full costume, including prosthetics and make-up took about three hours.

AFTERWORD

by Sir Percival Addlington —
Former Special Security Adviser to the Prime Minister

DOCTOR JOHN SMITH – FORMER SCIENTIFIC ADVISER TO UNIT. NOT THAT THEY EVER LISTENED MUCH.

Yes, there actually was an 'Afterword'. I don't know why – I mean, no one ever reads them, do they? So – and you'll thank me for this – I got rid of it. What I have put here instead, while sneaking round the printers late one rather wet Wednesday night, is far more useful and interesting. With better grammar, probably. Though I wasn't really intending to add anything, just making sure that what the book tells you is actually useful and accurate. Which it is. Up to a point.

But you know, it's what it doesn't tell you that is most alarming. Because the monsters it describes are just so much more dangerous, terrifying and (it goes without saying) monstrous than that.

And what about the monsters who don't get a mention? Like the Quarks – deadly robot servants of the cruel Dominators... Or the Cushonians, which are especially unpleasant as they live under your sofa and lie in wait for anyone who tries to hide behind it. Or the Axons, or the terrible Zodin...

So really what I'm saying is – this book gives you a good start. But don't imagine it tells the whole story. It just scratches the surface with its claws. Keep your wits about you and your eyes open. Don't blink, don't get complacent.

And watch out for the monsters – because you never know when they might